VISUAL
DICTIONARY
ANIMALS

V I S U A L
D I C T I O N A R Y
ANIMALS

FOG CITY PRESS

Published by Fog City Press
814 Montgomery Street
San Francisco, CA 94133 USA

First printed 2004

ISBN 1 877019 71 2

Color reproduction by Colourscan Co Pte Ltd
Printed by Tien Wah Press Pte Ltd
Printed in Singapore

A Weldon Owen Production

FOG CITY PRESS
Chief Executive Officer: John Owen
President: Terry Newell
Publisher: Lynn Humphries
Creative Director: Sue Burk
Project Editor: Jessica Cox
Project Designer: Heather Menzies
Editorial Coordinator: Jennifer Losco
Consultants: George McKay, Richard Vogt,
    Chris MacDonald, Hugh Dingle
Production Manager: Caroline Webber
Production Coordinator: James Blackman
Sales Manager: Emily Jahn
Vice President International Sales: Stuart Laurence

Produced using arkiva retrieval technology
For further information, contact arkiva@weldonowen.com.au

Contents

REPTILES

338

AMPHIBIANS

396

FISHES

422

INVERTEBRATES

486

FACT FILE

552

Living World

WORLD CLIMATES

- Tropical
- Subtropical
- Desert and semi-desert
- Dry temperate
- Wet temperate
- Cold temperate
- Polar
- Mountain
- → Cool currents
- → Warm currents

Oceans
The oceans' climates change with latitude, and warm and cool currents.

Tropical rainforests
Tropical climates and rainforests are found in equatorial regions.

Subtropical savannas
Savanna grasslands are found across tropical and subtropical latitudes.

Deserts and semi-deserts
Many arid zones lie downwind of mountains, resulting in dry conditions.

DIVERSE ECOSYSTEMS

Planet Earth's weather systems have created
a range of habitats. Within each habitat is
a community of plants and animals, called
an ecosystem. To survive, individuals must
adapt to their environment, making the
most of the resources it has to offer.

Alpine and polar regions
These regions are typified
by their harsh climates
and barren landscapes.

Temperate grasslands
Dry temperate climates
are found in mid-latitudes,
and produce grasslands.

Temperate forests
Wet temperate climates
are found in mid-latitudes,
and produce forests.

Coniferous forests
Northern temperate (or
boreal) climates occur in
the northern hemisphere.

Changing Habitats

Carboniferous period
360–286 million years ago

Permian period
286–248 million years ago

Triassic period
248–208 million years ago

EVOLUTION OF HABITATS

Plants began to grow on Earth about 550
million years ago, creating the first environments
for animal life. At first spore-bearing mosses and
ferns dominated. Then seed-bearing conifers
and flowering plants emerged. As plants evolved
so did the animals that ate them, resulting in
the complex ecosystems of today.

Carboniferous
Horsetail fern

Permian
Ginko

Triassic
Cyclad

Jurassic
Conifer

Cretaceous
Magnolia

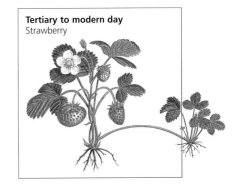

Tertiary to modern day
Strawberry

Jurassic period
208–144 million years ago

Cretaceous period
144–65 million years ago

Tertiary period
65 million years ago

Tropical Rainforests

BROMELIAD LIFE

Velvet worms can be found here, high above their usual forest-floor home.

Mosquitoes breed in the bromeliad pool.

Female poison frog deposits tadpoles into water-filled bromeliads.

Alligator lizard drinks from the pool, safe above the forest floor.

Beetle larvae prey on poison frog tadpoles.

Land snails favor the moist leaves for foraging and hiding.

Mouse opossum eats a damselfly larva from the bromeliad pool.

RAINFOREST PLANTS

Leaf dripping rainwater

Flowering ginger

Epiphytic plant

Savannas

SAVANNA PLANTS

Sturt's pea

Wheat

Eucalyptus flower

GELADA BABOON HABITAT

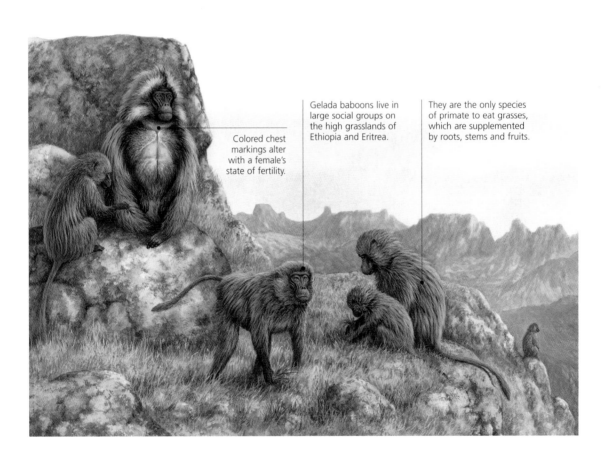

Colored chest markings alter with a female's state of fertility.

Gelada baboons live in large social groups on the high grasslands of Ethiopia and Eritrea.

They are the only species of primate to eat grasses, which are supplemented by roots, stems and fruits.

SUBTROPICAL CLIMATE
Subtropical areas generally receive rainfall only in summer. This provides perfect conditions for savanna grasslands rather than dense forests.

Deserts

DESERT BY DAY

Plant and animal species that live in deserts have adapted to the lack of water. Some plants store water in their stems; others have long roots reaching underground.

DESERT LIFE

Jeweled gecko feeds on invertebrates in the spinifex, emitting a toxic fluid when startled.

Crimson chats live in open plains with a cover of spinifex or saltbush.

Small nocturnal cockroaches feed on decaying organic matter.

Ningauis are tiny marsupials that quench their thirst by licking dew from spinifex.

Burton's legless lizard preys on smaller lizards, suffocating and then swallowing them.

Desert skink eats termites and other invertebrates in the spinifex community.

DESERT BY NIGHT

Because of the intense heat of the day in some regions, many desert animals are nocturnal. Some plants have adapted to flower at night so they will be pollinated.

Beaver-tail cactus

Quiver tree

Century plant flowers

Chain-fruit cholla

Temperate Regions

TEMPERATE CLIMATE

Mid-latitude areas are typified by four distinct seasons. Depending on rainfall, temperate regions can be grasslands, shrublands or woodlands. To survive cold winters, animals may migrate or hibernate.

Spring: leaves sprout; flowers bloom.

FOUR SEASONS

Summer: fertilized flowers form fruits.

Autumn: fruits seed; leaves fall.

Winter: buds protect new growth.

TEMPERATE LIFE

Pika

Cuckoo

Cattle

Woodlands

WOODLAND PLANTS

Autumn leaves

Pine cones

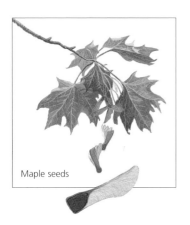

Maple seeds

CONIFEROUS FORESTS

Evergreen conifers thrive in the forests of cold temperate regions. The trees are cone-shaped, allowing snow to slide off. Many animals have thick fur for warmth, while some hibernate to escape the bitter cold.

Gamebirds such as grouse have thick, downy plumage to survive the cold winters.

With coloring like mottled bark, owls are well suited to coniferous forests.

Wolverines are skilled hunters in deep snow. Their fur is hollow for insulation.

Solitary mooses, the largest of all deer, gather in groups only in the breeding season.

To survive winter, squirrels gather a cache of nuts and berries every autumn.

Alpine Regions

ALPINE PLANTS

Edelweiss

Puya

Conifers

HYRAX HABITAT

Rock and bush hyraxes live in close proximity.

Rock hyraxes feed on alpine grasses.

Bush hyraxes forage for leaves and bark.

ALPINE CLIMATE

As altitudes and latitudes increase, vegetation decreases, until trees can no longer grow. Mountain areas are similar to arctic tundra: plants and animals adapt to flourish in short summers and survive long winters.

Seashores

COASTAL PLANTS

Coconut sprouting

Screw pine fruit

Casuarina leaves

SEASHORES

Only salt-resistant plants can grow close to the sea, with grasses and shrubs forming the frontline against sea breezes. Below the splash line, tidal plants and animals thrive in the pools and among rocks.

Hermit crabs protect their soft bodies in discarded mollusk shells.

Mussels grow in dense colonies attached to rocks.

Clinging to rocks like mollusks, barnacles are actually crustaceans.

Blennies hide under stones, darting from crevice to shadow.

Carnivorous starfish move slowly on hundreds of feet.

Small octopuses are secretive inhabitants of tidepools.

Seaweeds are marine algae, and lack a plant's roots, flowers, seeds and fruit.

Colorful sea anemones feed on small animals stung by their tentacles.

Sculpin are predators that lie in wait for prey rather than chase it.

Crabs are abundant in intertidal tidepools.

Oceans

OCEAN HABITATS

The oceans contain many habitats, dependent on latitude and depth. A huge variety of animals and plants have adapted to life underwater, from seaweeds and seagrasses, to sharks, corals and fishes.

Dugongs live among the coastal seagrass beds in the Indo-Pacific region.

Yellow-finned leatherjackets are herbivorous fishes.

Blue swimmer-crabs bury themselves in sand up to their eyes to wait for prey.

Fan-mussels anchor into the sand with only part of their shell exposed.

OCEAN LIFE

0

Sunlight zone
Enough sunlight
penetrates this layer for
plants to live. It contains
most of the ocean's life.

650 ft
(200 m)

Solitary green turtles
migrate vast distances
to lay their eggs.

Twilight zone
Not enough sunlight to
sustain plant life. Many
fishes in this zone contain
light-producing bacteria
and glow in the dark.

3,250 ft
(1,000 m)

Midnight zone
This is a vast mass of
cold, slow-moving water
in absolute darkness. Life
is about a tenth that of
the twilight zone and is
also less diverse.

9,850 ft
(3,000 m)

Abyssal zone
The near-freezing, pitch-
black bottom layer has
few species and little food.
Near hydrothermal vents,
some animals make food
using internal bacteria
and hydrogen sulfide.

Web of Life

First single-celled organisms

Atmosphere forms.

Earth forms.

4 bya

First simple plants (cyanobacteria)

3.5 bya

3 bya

Single-celled organisms with a nucleus form.

2.5 bya

First multi-cellular organisms (animals)

2 bya

1.5 bya

First sexual reproduction

1 bya

505 mya

Jellyfish and seaweeds dominate the oceans.

Trilobites and other hard-shelled marine invertebrates form the 'Cambrian explosion.'

CAMBRIAN

550 mya

First reptiles; insects with wings evolve.

Mammal-like reptiles

PERMIAN

CARBONIFEROUS

286 mya

360 mya

First dinosaurs and mammals

248 mya

First marsupials and placental mammals

TRIASSIC

Dinosaurs diversify; first birds evolve.

208 mya

JURASSIC

144 mya

FOSSIL RECORD

Trilobite

Heliobatus ray

Snakefly

Dinosaur embryo

▢	Archean era	▢	Cenozoic era
▢	Proterozoic era	▬	Mass extinction
▢	Paleozoic era	**bya**	billion years ago
▢	Mesozoic era	**mya**	million years ago

ORDOVICIAN

435 mya — First fishes and corals evolve.

Jawed fishes; first land plants and animals

SILURIAN

Fishes dominate; first insects and amphibians, first plants with seeds

DEVONIAN

408 mya

Large mammals and primates evolve.

Early horses, camels, rodents, elephants, whales and bats

Large browsing mammals; apes, dogs and bears

Humans (*Homo sapiens*) evolve.

CRETACEOUS

65 mya

TERTIARY

QUATERNARY

2 mya   Present

Before the Dinosaurs

Ornithosuchus   Dimetrodon   Dragonfly   Hylonomus   Ichthyostega   Dunkleosteus

Blue-green algae

Jellyfish

Ammonite

Trilobite

Pteraspis

Drepanaspis

Scorpion

PRE-DINOSAUR ERAS

Archaean

Proterozoic

Cambrian

Ordovician

Silurian

Devonian

Carboniferous

Permian

Triassic Period

Triassic landmasses

TRIASSIC LANDSCAPE

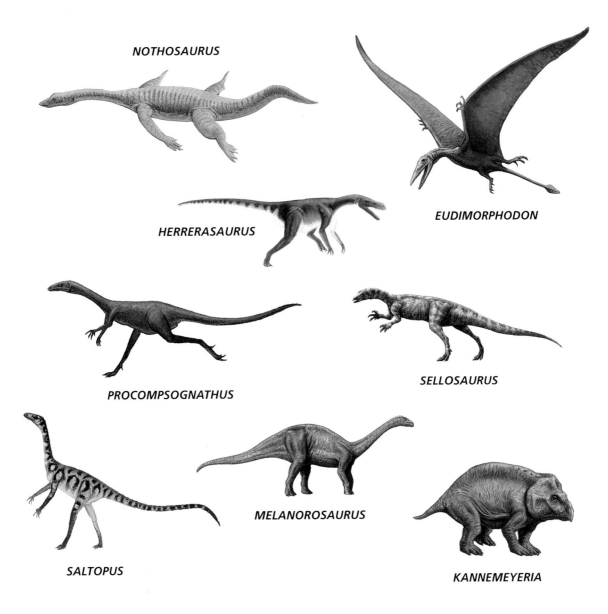

NOTHOSAURUS

EUDIMORPHODON

HERRERASAURUS

PROCOMPSOGNATHUS

SELLOSAURUS

SALTOPUS

MELANOROSAURUS

KANNEMEYERIA

TRIASSIC DINOSAUR DIETS

Dragonfly

Wielandiella

Haramiya

Tree fern

Jurassic Period

Jurassic landmasses

JURASSIC LANDSCAPE

STEGOSAURUS

ALLOSAURUS

CAMPTOSAURUS

COELURUS

CERATOSAURUS

RHAMPHORHYNCHUS

HETERODONTOSAURUS

DIPLODOCUS

JURASSIC DINOSAUR DIETS

Pleisochelys

Archaeopteryx

Cockroach

Cretaceous Period

Cretaceous landmasses

CRETACEOUS LANDSCAPE

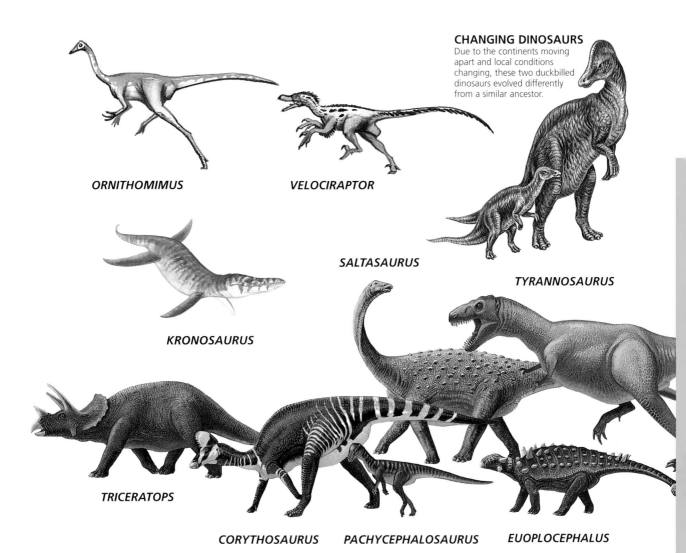

ORNITHOMIMUS

VELOCIRAPTOR

CHANGING DINOSAURS
Due to the continents moving apart and local conditions changing, these two duckbilled dinosaurs evolved differently from a similar ancestor.

SALTASAURUS

TYRANNOSAURUS

KRONOSAURUS

TRICERATOPS

CORYTHOSAURUS

PACHYCEPHALOSAURUS

EUOPLOCEPHALUS

CRETACEOUS LIFE

Polyglyphanodon

Magnolia

Crusafontia

Types of Dinosaurs

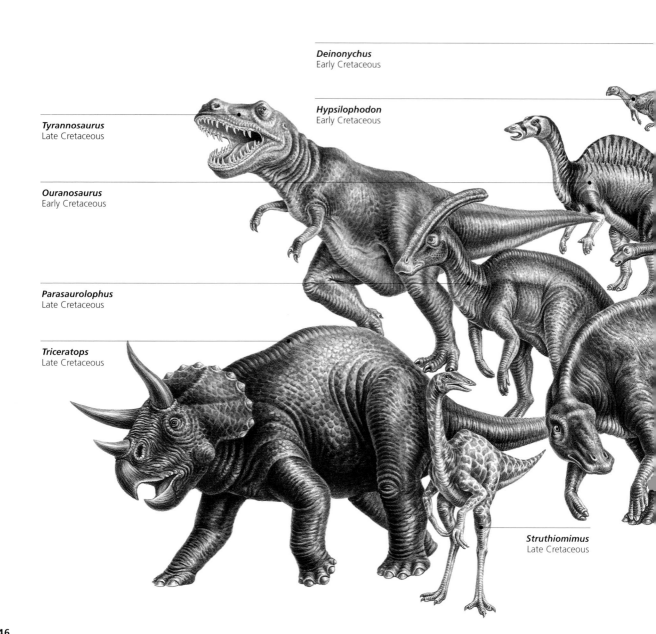

Deinonychus
Early Cretaceous

Tyrannosaurus
Late Cretaceous

Hypsilophodon
Early Cretaceous

Ouranosaurus
Early Cretaceous

Parasaurolophus
Late Cretaceous

Triceratops
Late Cretaceous

Struthiomimus
Late Cretaceous

Brachiosaurus
Late Jurassic

Allosaurus
Late Jurassic/
early Cretaceous

Coelurus
Late Jurassic

Plateosaurus
Late Triassic

Stegosaurus
Late Jurassic

Coelophysis
Late Triassic

Euoplocephalus
Late Cretaceous

Saltasaurus
Late Cretaceous

Pachycephalosaurus
Late Cretaceous

Maiasaura
Late Cretaceous

Dinosaur Characteristics

DINOSAUR ANATOMY

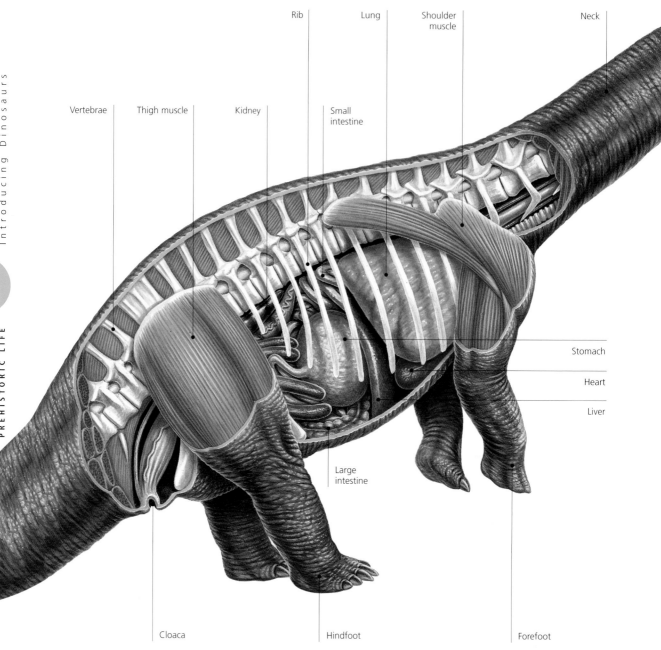

Rib

Lung

Shoulder
muscle

Neck

Vertebrae

Thigh muscle

Kidney

Small
intestine

Stomach

Heart

Liver

Large
intestine

Cloaca

Hindfoot

Forefoot

Head

Tyrannosaurus

Troödon

Stegosaurus

Human

TYPES OF TEETH

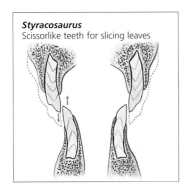

Styracosaurus
Scissorlike teeth for slicing leaves

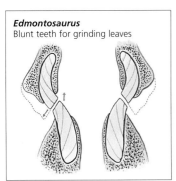

Edmontosaurus
Blunt teeth for grinding leaves

49

Dinosaur Hips

EVOLUTION OF HIPS

Legs sprawling on either side of body

Legs partially under body

Legs fully under body

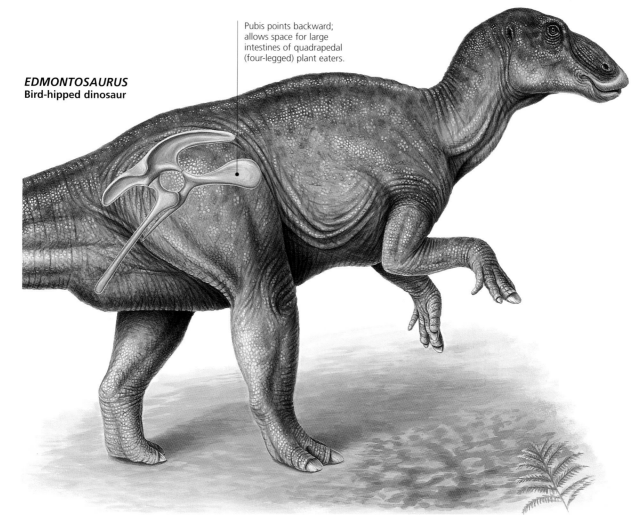

Pubis points backward; allows space for large intestines of quadrapedal (four-legged) plant eaters.

EDMONTOSAURUS
Bird-hipped dinosaur

TWO TYPES OF HIPS
Even though ornithischian dinosaurs are called 'bird-hipped,' birds actually evolved from saurischian or 'lizard-hipped' dinosaurs.

Ornithischian pelvic bones

Illium

Ischium

Pubis

Saurischian pelvic bones

Illium supports leg muscles.

Pubis supports legs.

Ischium supports tail.

Femur

Tibia

Ankle

ALLOSAURUS
Lizard-hipped dinosaur

Pubis points forward between the legs; pelvic bones form a rigid triangle to support fast running.

Raising Young

HATCHING OUT

Maiasaura hatchling

One-year-old Maiasaura

Adult Maiasaura

INSIDE AN EGG

Allantois stored waste.

Membrane (chorion) provided oxygen.

Albumen (yolk sac) nourished embryo.

Amniotic (fluid) sac cushioned dinosaur.

Eggshell protected embryo.

OVIRAPTOR NEST

Mother *Oviraptor* feeds a freshly killed young *Velociraptor* to her nest of hatchlings. Unlike most modern birds and reptiles, some dinosaurs cared for their young after hatching.

DINOSAUR NEST

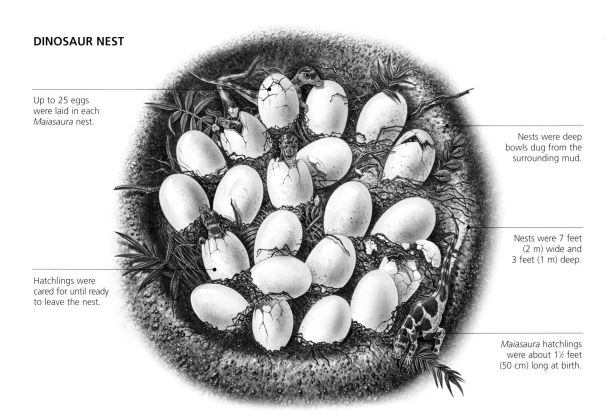

Up to 25 eggs were laid in each *Maiasaura* nest.

Nests were deep bowls dug from the surrounding mud.

Nests were 7 feet (2 m) wide and 3 feet (1 m) deep.

Hatchlings were cared for until ready to leave the nest.

Maiasaura hatchlings were about 1½ feet (50 cm) long at birth.

EGG ARRANGEMENTS

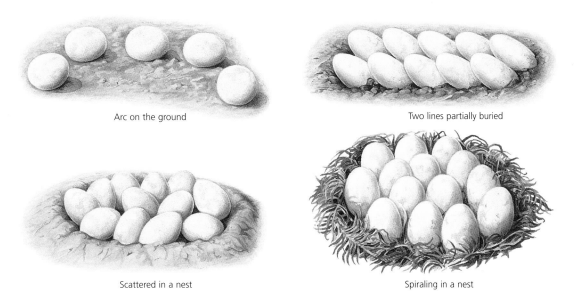

Arc on the ground

Two lines partially buried

Scattered in a nest

Spiraling in a nest

Sauropods

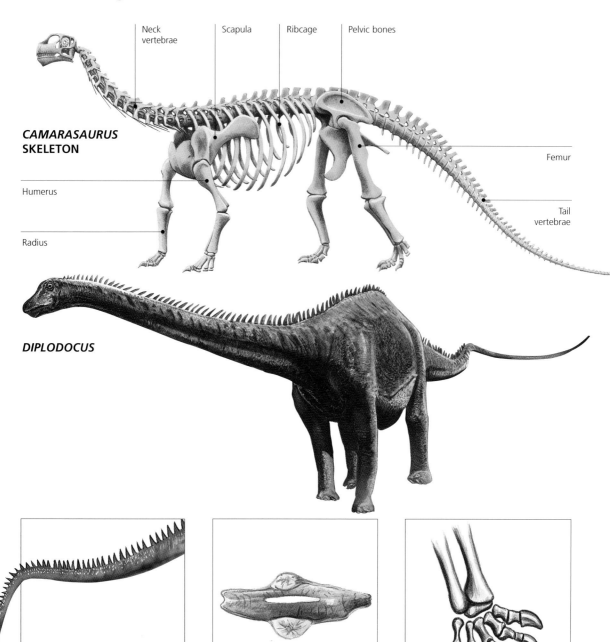

Neck vertebrae

Scapula

Ribcage

Pelvic bones

CAMARASAURUS SKELETON

Humerus

Radius

Femur

Tail vertebrae

DIPLODOCUS

Diplodocus tail

Diplodocus tail vertebra

Diplodocus foot bones

PLATEOSAURUS

Plateosaurus skull

BRACHIOSAURUS

Brachiosaurus head

Brachiosaurus skull

Brachiosaurus head and neck

Theropods

STRUTHIOMIMUS SKELETON

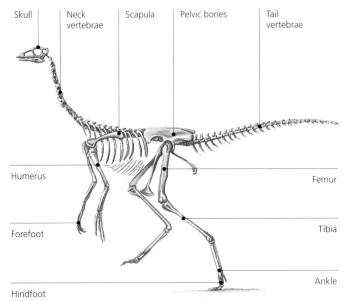

Skull

Neck
vertebrae

Scapula

Pelvic bones

Tail
vertebrae

Humerus

Femur

Forefoot

Tibia

Ankle

Hindfoot

EATING TOOLS

Allosaurus skull
Hinged jaws to swallow prey whole

Carcharodontosaurus teeth
Serrated teeth front and back to slice flesh

STRUTHIOMIMUS

ADAPTATIONS FOR SURVIVAL

Tyrannosaurus had hinged jaws and backward-facing teeth to rip and tear flesh.

Gallimimus ran at speeds of up to 30 miles (48 km) an hour. A slower dinosaur such as *Albertosaurus* had no chance of catching it.

FEEDING

Baryonyx

Oviraptor

Albertosaurus

Compsognathus

Theropods

VELOCIRAPTOR

Velociraptor arm bones

Velociraptor foot

MEGALOSAURUS

Megalosaurus tooth

Serrated edge

ALLOSAURUS

Allosaurus skull

OVIRAPTOR

Two Oviraptor heads showing crests

Ornithopods

OURANOSAURUS SKELETON

Head

Neck vertebrae

Forelimb

Pelvic bones

Femur

Hindlimb

Extended upper
vertebrae

Ribcage

Lower
vertebrae

Ankle

ORNITHOPOD SKULLS

Ouranosaurus skull

Anatotitan skull

Lambeosaurus skull

Corythosaurus skull

OURANOSAURUS

IGUANODONS

PARASAUROLOPHUS

Juvenile

Adult female

Adult male

Armored Dinosaurs

STEGOSAURUS ANATOMY

Plates along spine may have been used to regulate temperature.

Head

Neck vertebrae

Ribcage

Lower vertebrae

Tail spikes for defense

Scutellosaurus skin detail

Ankylosaurus tail club

STEGOSAURUS

STEGOSAURUS PLATES

Tail spikes

Lower back plates

Dorsal plates

TUOJIANGOSAURUS

Tuojiangosaurus tail
Spikes used to stab
potential predators

EUOPLACEPHALUS

Euoplacephalus tail
Clublike bony tail
used to hit predators

Euoplacephalus skull
Armor-plated with
spikes to protect neck

EDMONTONIA

Edmontonia head
Forward-facing spikes
used to ram predators

Ceratopians

CHASMOSAURUS HEAD

CHASMOSAURUS

STYRACOSAURUS SKULL

STYRACOSAURUS

TRICERATOPS

PROTOCERATOPS

PROTOCERATOPS SKULL

TOROSAURUS SKULL

TOROSAURUS

CENTROSAURUS

CENTROSAURUS HEAD

65

Dinosaur Contemporaries

DIMORPHODON

PTERODAUSTRO

ARCHAEOPTERYX

DIMETRODON

MESONYX

ALPHADON

DORUDON

Peloneustes
Pliosaur
10 feet (3 m)

Archelon
Turtle
13 feet (4 m)

ICHTHYOSAURUS

Nothosaurus
Nothosaur
10 feet (3 m)

Platecarpus
Lizard
13 feet (4 m)

LIOPLEURODON

Deinosuchus
Crocodilian
49 feet (15 m)

67

Dinosaur Extinction

EXTINCTION THEORIES

Hot climate change

Cold climate change

Volcanic eruptions

Meteorite crashing to Earth

DINOSAUR EXTINCTION

There are several theories about what caused the dinosaurs to suddenly disappear 65 million years ago. The most widely accepted theory is that a huge meteorite hit Earth causing environmental chaos.

Small mammal

AGE OF MAMMALS
The small mammals of the Cretaceous were some of the species that survived the mass extinctions of 65 million years ago. These evolved and diversified into the thousands of known species today.

Recreating the Dinosaurs

FOSSIL FORMATION

Safe from scavengers
A dead dinosaur lies decomposing at the bottom of a lake, its skeleton remaining intact.

Covered up
Layers of sand or silt cover the dinosaur's bones, which stops them from being washed away.

Turning to stone
Trapped and flattened by layers of sedimentary rock, dinosaur bones are replaced by minerals.

Fossil uncovered
Millions of years later, movements within the Earth bring the fossilized skeleton to the surface.

Pieces of fossilized tail vertebrae

Missing sections of vertebrae are sketched from surrounding bones.

Sketches are painstakingly executed.

WALKING ON THE PAST

DINOSAUR BONES

Tail vertebrae

Scapula

Hip bone

Femur

Neck vertebra

Rib

RECONSTRUCTING A DINOSAUR

Muscle layers are added by studying existing reptiles.

By looking closely at bones, small muscles are created.

Final layers of flesh and skin are added, through guesswork.

COLOR AND TEXTURE
Because no one knows what dinosaurs actually looked like, illustrators have to guess what the color and texture of dinosaur skin was like.

Modern Relatives

ARCHAEOPTERYX

It is generally believed that modern birds evolved from dinosaurs. The skeleton of *Archaeopteryx*, the earliest known bird, is very similar to certain dinosaurs—such that many now classify *Archaeopteryx* as a dinosaur with feathers.

FROM DINOSAUR TO BIRD

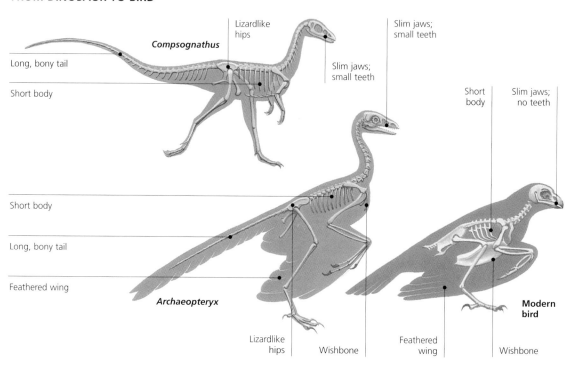

Compsognathus

Lizardlike hips

Long, bony tail

Short body

Slim jaws; small teeth

Slim jaws; small teeth

Short body

Slim jaws; no teeth

Short body

Long, bony tail

Feathered wing

Archaeopteryx

Lizardlike hips

Wishbone

Feathered wing

Modern bird

Wishbone

Waterfowl

Turtle

Crocodile

Fish

Emu

Seabird

Hoatzin

Lizard

Insect

Archaeopteryx wing

Bat wing

Pigeon wing

73

Classifying Mammals

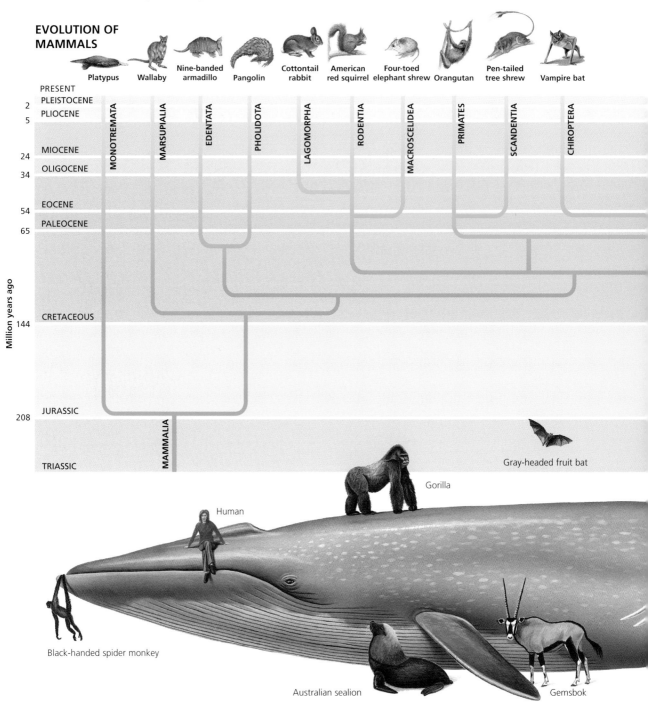

EVOLUTION OF MAMMALS

Platypus · Wallaby · Nine-banded armadillo · Pangolin · Cottontail rabbit · American red squirrel · Four-toed elephant shrew · Orangutan · Pen-tailed tree shrew · Vampire bat

Million years ago

PRESENT
PLEISTOCENE
2
PLIOCENE
5
MIOCENE
24
OLIGOCENE
34
EOCENE
54
PALEOCENE
65
CRETACEOUS
144
JURASSIC
208
TRIASSIC

MONOTREMATA · MARSUPIALIA · EDENTATA · PHOLIDOTA · LAGOMORPHA · RODENTIA · MACROSCELIDEA · PRIMATES · SCANDENTIA · CHIROPTERA

MAMMALIA

Introducing Mammals

MAMMALS

Gray-headed fruit bat

Gorilla

Human

Black-handed spider monkey

Australian sealion

Gemsbok

Philippine flying lemur

Greater moonrat

Manchurian tiger

Fallow deer

Humpback whale

Aardvark

Arabian horse

Rock hyrax

Manatee

African elephants

DERMOPTERA

INSECTIVORA

CARNIVORA

ARTIODACTYLA

CETACEA

TUBULIDENTATA

PERISSODACTYLA

HYRACOIDEA

SIRENIA

PROBOSCIDEA

Blue whale

Giraffe

African elephant

Black rhinoceros

Beaver

Mammal Characteristics

HUNTERS AND PREY

Forward-facing eyes for sight in three dimensions

Sideways-facing eyes for 180° vision

Heavily muscled hindlimbs for rapid acceleration

Claws to grasp prey

Slender hindlimbs for lightness with power

Toes modified to form hooves for greater leverage when running

MAMMAL FOREFEET

Whale: modified flipper for swimming

Bear: adapted for digging and hunting

Bat: fingers stretched for flying

GIVING BIRTH TO LIVE YOUNG

All mammals are vivaporous: they give birth to live young.

Wildebeest birth
Social mammals time mating so their young are born together, to protect them from predators.

After birth
Newly born wildebeests are able to stand and run within minutes of birth.

MAMMAL EAR

The three bones that make up a mammal's middle ear work together to transmit sound waves into the sensory cells of the inner ear.

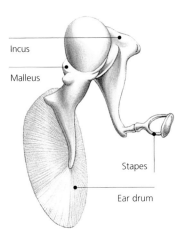

Incus

Malleus

Stapes

Ear drum

CONVERGENT EVOLUTION

Animal species that are not closely related often adapt to environmental conditions or pressures in similar ways. This is called convergent evolution.

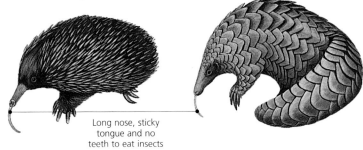

Long nose, sticky tongue and no teeth to eat insects

Short-beaked echidna

Pangolin

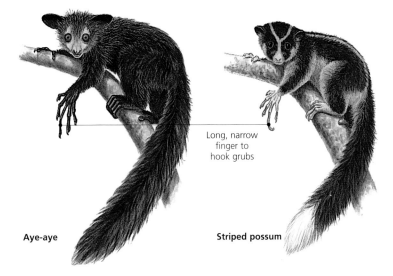

Long, narrow finger to hook grubs

Aye-aye

Striped possum

MAMMAL HAIR

All mammals have hair on their bodies. A typical hair contains three layers: an insulating medulla, cortex containing pigment and outer cuticle.

Cortex

Medulla

Cuticle

Slow moving to conserve energy

Koala

Sloth

Evolution of Mammals

DIMETRODON
Mammal-like reptile that
lived 300 million years ago.

MEGAZOSTRODON
Earliest known true mammal,
it lived 220 million years ago.

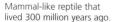

CYNOGNATHUS
Mammal-like reptiles that lived
245–230 million years ago.

MORGANUCODONTID
One of the earliest true mammals,
it lived 195 million years ago.

UINTATHERIUM
One of the earliest herbivores,
it lived 60 million years ago.

ARSINOITHERIUM
Relative of modern hyraxes,
it lived 40 million years ago.

INDRICOTHERIUM
The largest land mammal ever
lived 30 million years ago.

URSAVUS
Thought to be the ancestor of modern
bears, it lived 20 million years ago.

WOOLLY MAMMOTHS
Ice-age mammals related to modern elephants
that lived 350–10 thousand years ago.

Monotremes

Location

- Echidnas
- Echidnas and platypus

ECHIDNA ADAPTATIONS

Claws strong enough to break into termite nests

Long, sticky tongue

Termites, ants or grubs

SHORT-BEAKED ECHIDNA
Tachyglossus aculeatus

LONG-BEAKED ECHIDNA
Zaglossus bruijni

WATERTIGHT EYES

Eyes open when out of water

Eyes closed when diving

PLATYPUS
Ornithorhynchus anatinus

PLATYPUS NEST

Monotremes are the only mammals that lay eggs and do not give birth to live young.

CLAWS FOR DIGGING

Web on forefoot drawn back

Web on forefoot extended

Marsupials

Location

- American marsupials
- Australasian marsupials

NUMBAT
Myrmecobius fasciatus

REPRODUCTIVE ORGANS

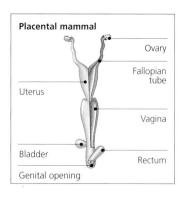

Placental mammal

- Ovary
- Fallopian tube
- Uterus
- Vagina
- Bladder
- Rectum
- Genital opening

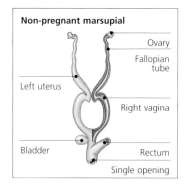

Non-pregnant marsupial

- Ovary
- Fallopian tube
- Left uterus
- Right vagina
- Bladder
- Rectum
- Single opening

Pregnant marsupial

- Ovary
- Fallopian tube
- Left uterus
- Median vagina develops for birthing.
- Left vagina
- Single opening

STARTING OUT

Embryonic young leaves birth canal.

It makes its way to its mother's pouch.

Finds nipple inside the pouch and suckles.

JUMPING INTO THE POUCH

The young kangaroo pushes front
feet and head into the pouch.

It rolls forward into the pouch.

Then twists around so it can see
and jump out of the pouch.

RED KANGAROO
Macropus rufus

STRIPED POSSUM
Dactylopsila trivirgata

TASMANIAN DEVIL
Sarcophilus laniarius

Marsupials

SPOTTED CUSCUS
Spilocuscus sp.

COMMON WOMBAT
Vombatus ursinus

SQUIRREL GLIDER
Petaurus norfolcensis

WATER OPOSSUM
Chironectes minimus

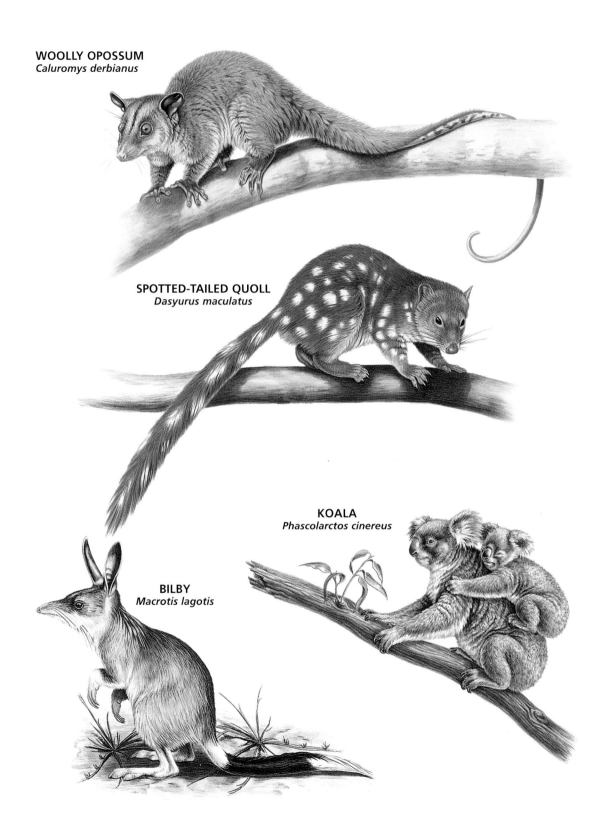

WOOLLY OPOSSUM
Caluromys derbianus

SPOTTED-TAILED QUOLL
Dasyurus maculatus

KOALA
Phascolarctos cinereus

BILBY
Macrotis lagotis

Kangaroos and Wallabies

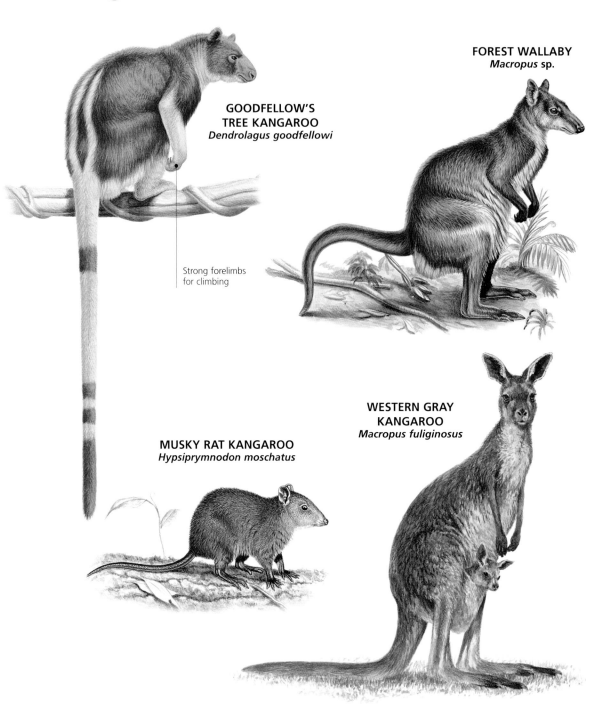

**GOODFELLOW'S
TREE KANGAROO**
Dendrolagus goodfellowi

FOREST WALLABY
Macropus sp.

Strong forelimbs
for climbing

MUSKY RAT KANGAROO
Hypsiprymnodon moschatus

**WESTERN GRAY
KANGAROO**
Macropus fuliginosus

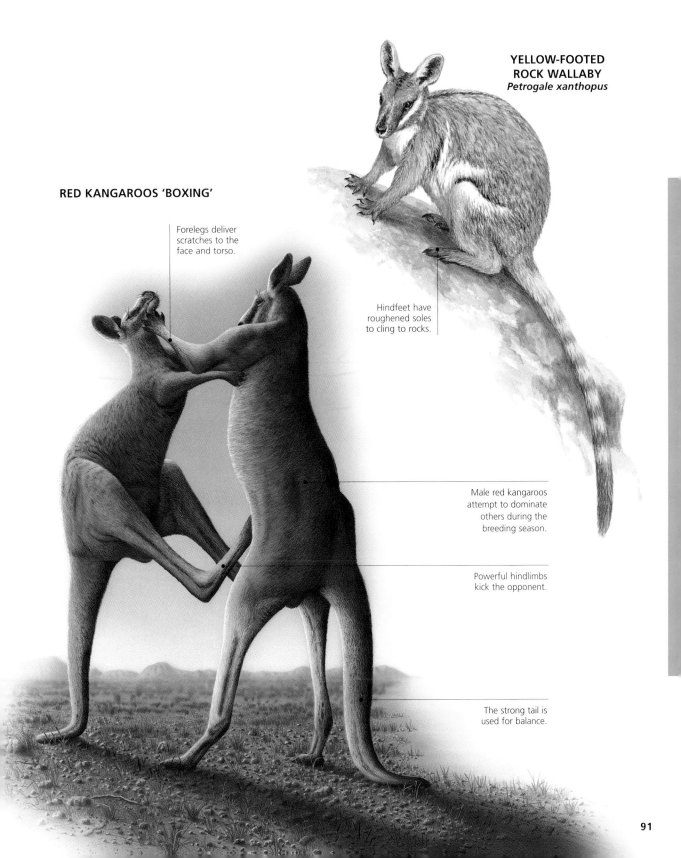

YELLOW-FOOTED ROCK WALLABY
Petrogale xanthopus

RED KANGAROOS 'BOXING'

Forelegs deliver scratches to the face and torso.

Hindfeet have roughened soles to cling to rocks.

Male red kangaroos attempt to dominate others during the breeding season.

Powerful hindlimbs kick the opponent.

The strong tail is used for balance.

91

Xenarthrans and Pangolins

Location

■ Xenarthrans (anteaters, armadillos and sloths)
■ Pangolins

TAMANDUA
Tamandua sp.

MANED THREE-TOED SLOTH
Bradypus torquatus

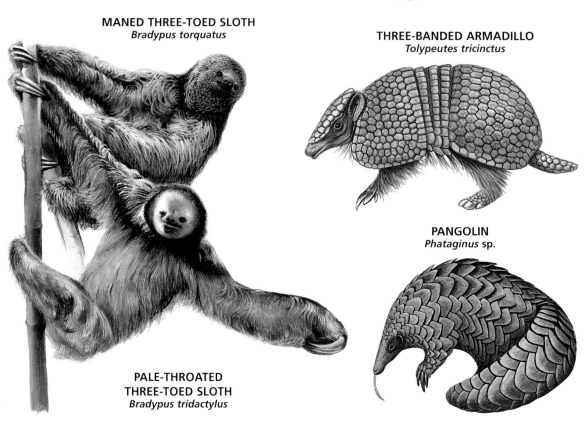

**PALE-THROATED
THREE-TOED SLOTH**
Bradypus tridactylus

THREE-BANDED ARMADILLO
Tolypeutes tricinctus

PANGOLIN
Phataginus sp.

GIANT ANTEATER
Myrmecophaga tridactyla

Baby anteaters are carried on their mother's back.

Long bushy tail is covered with coarse fur.

Tongue can be more than two feet (60 cm) long.

Tubelike snout has no teeth.

It walks on its knuckles to protect claws from wear.

Termites are the anteater's main diet.

Insect-eating Mammals

Location

■ Insect-eating mammals

WESTERN EUROPEAN HEDGEHOG
Erinaceus europaeus

TYPES OF NOSES

Algerian hedgehog
Short pointed snout
with bristles

European mole
Sensitive nose to smell
and feel for prey

Pyrenean desman
Flexible probe to find
insects underwater

MOONRAT
Echinosorex gymnura

PYRENEAN DESMAN
Galemys pyrenaicus

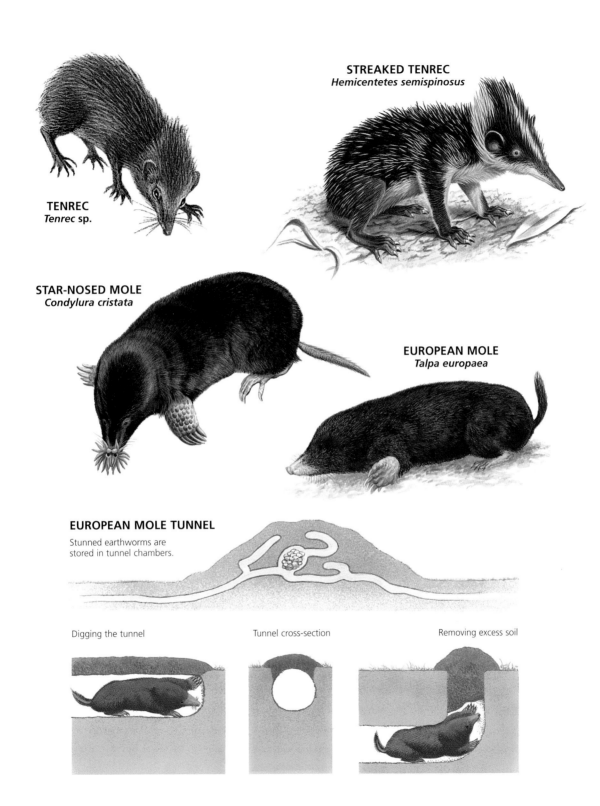

TENREC
Tenrec sp.

STREAKED TENREC
Hemicentetes semispinosus

STAR-NOSED MOLE
Condylura cristata

EUROPEAN MOLE
Talpa europaea

EUROPEAN MOLE TUNNEL

Stunned earthworms are
stored in tunnel chambers.

Digging the tunnel

Tunnel cross-section

Removing excess soil

Flying Lemurs and Tree Shrews

Location

- Flying lemurs
- Flying lemurs and tree shrews

MALAYAN FLYING LEMUR
Cynocephalus variegatus

PHILIPPINE FLYING LEMUR
Cynocephalus volans

ANATOMY OF GLIDING

Patagium joins
body at neck.

Elongated
forelimb

Patagium
stretches between
limbs and toes.

Long, slender fingers
have strong claws
for climbing.

Kite-shaped
membrane
(patagium)

Underbelly

Tail

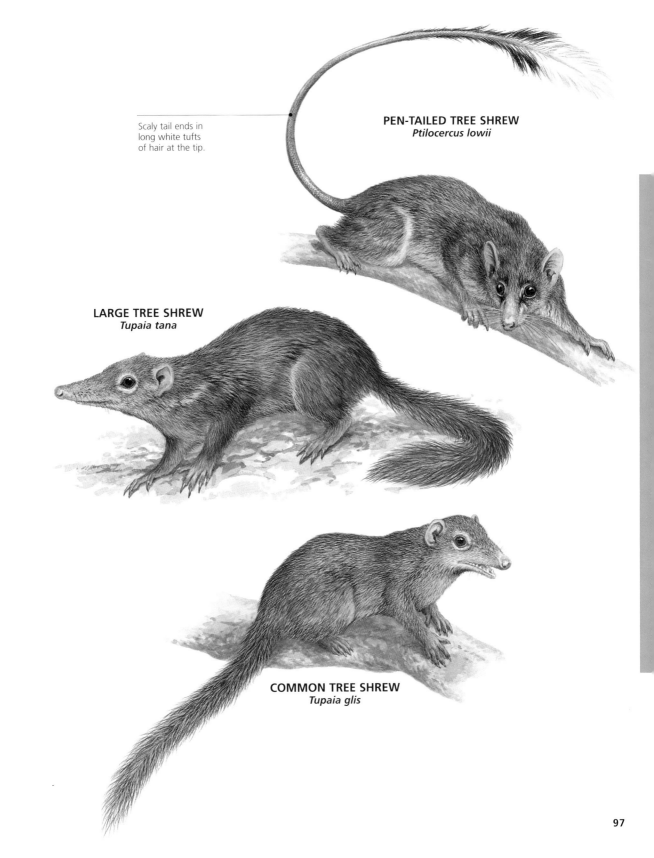

Scaly tail ends in long white tufts of hair at the tip.

PEN-TAILED TREE SHREW
Ptilocercus lowii

LARGE TREE SHREW
Tupaia tana

COMMON TREE SHREW
Tupaia glis

Bats

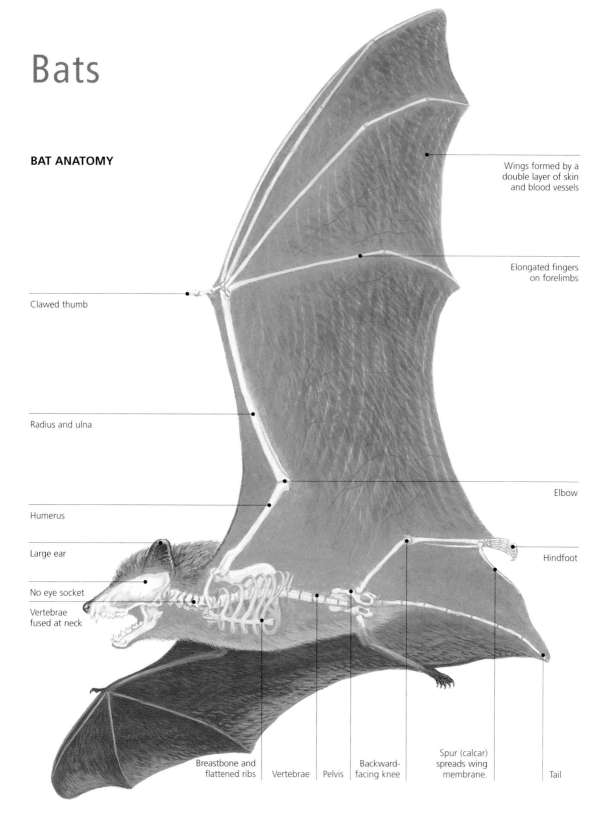

BAT ANATOMY

Wings formed by a
double layer of skin
and blood vessels

Elongated fingers
on forelimbs

Clawed thumb

Radius and ulna

Elbow

Humerus

Large ear

Hindfoot

No eye socket

Vertebrae
fused at neck

Breastbone and
flattened ribs

Vertebrae

Pelvis

Backward-
facing knee

Spur (calcar)
spreads wing
membrane.

Tail

ECHOLOCATION

Insect-eating bats use pulses of ultrasound to detect and capture prey.

Bat emits rapid 'clicks' of sound.

The 'clicks' bounce off prey back to bat, revealing its location.

Frequency of 'clicks' increases as bat nears to pinpoint prey.

Bat seizes prey.

VARIATIONS

Trident bat

Ear detects reflected sound.

Nose emits 'clicks' of sound.

Whiskered bat

Ear detects reflected sound.

Mouth emits 'clicks' of sound.

EARS AND NOSES

Long-eared bat
Large ears with flaps may improve accuracy of echolocation.

Lesser bare-backed fruit bat
Tube-shaped nostrils sniff out food.

Tent-building bat
Flaps of skin on leaf-shaped nose help detect echoes from prey.

Old World Fruit Bats

Location

■ Old World fruit bats

**GAMBIAN EPAULETTED
FRUIT BAT**
Epomorphorus gambianus

HAMMER-HEADED FRUIT BAT
Hypsignathus monstrosus

GRAY-HEADED FLYING FOX
Pteropus poliocephalus

Bats

MAMMALS

100

EGYPTIAN FRUIT BAT
Rousettus egyptiacus

FRUIT BAT EATING
Many fruit bats eat upside down, clutching their meal in their claws. Their relatively well-developed second digit and claw aids in this process.

INDIAN FLYING FOX
Pteropus giganteus

Insect-eating Bats

Location

■ Insect-eating bats

VAMPIRE BAT
Desmodus rotundus

**AFRICAN
YELLOW-WINGED BAT**
Lavia frons

HORSESHOE BAT
Rhinolophus sp.

SPOTTED BAT
Euderma maculatum

GREATER BULLDOG BAT
Noctilio leporinus

HONDURAN WHITE BATS
Ectophylla alba
Honduran white bats cut away
the connection between the
midrib and edge of palm
fronds, causing the fronds to
curl. They then roost in the
'tent' they have created.

WRINKLE-FACED BAT
Centurio senex

TENT-BUILDING BAT
Uroderma bilobatum

SWORD-NOSED BAT
Lonchorhina sp.

Primates

EVOLUTION OF PRIMATES

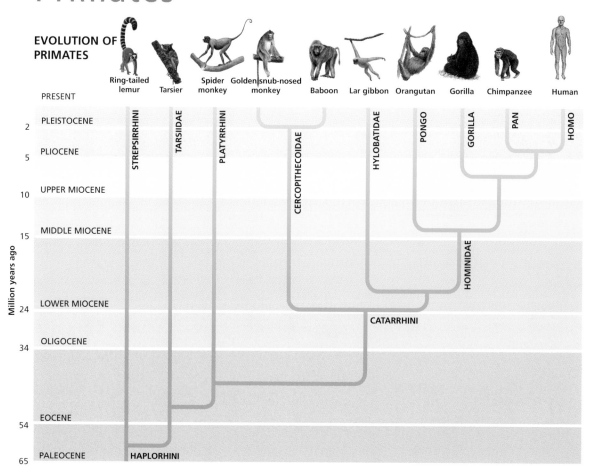

Ring-tailed lemur | Tarsier | Spider monkey | Golden snub-nosed monkey | Baboon | Lar gibbon | Orangutan | Gorilla | Chimpanzee | Human

PRESENT

Million years ago	
2	PLEISTOCENE
5	PLIOCENE
10	UPPER MIOCENE
15	MIDDLE MIOCENE
24	LOWER MIOCENE
34	OLIGOCENE
54	EOCENE
65	PALEOCENE

STREPSIRRHINI

TARSIIDAE

PLATYRRHINI

CERCOPITHECOIDAE

HYLOBATIDAE

PONGO

GORILLA

PAN

HOMO

HOMINIDAE

CATARRHINI

HAPLORHINI

PRIMATE FEET AND HANDS

Indri: strong, wide grip

Aye-aye: uses claws to climb

Tarsier: disklike pads on toes

Gorilla: broad to support weight

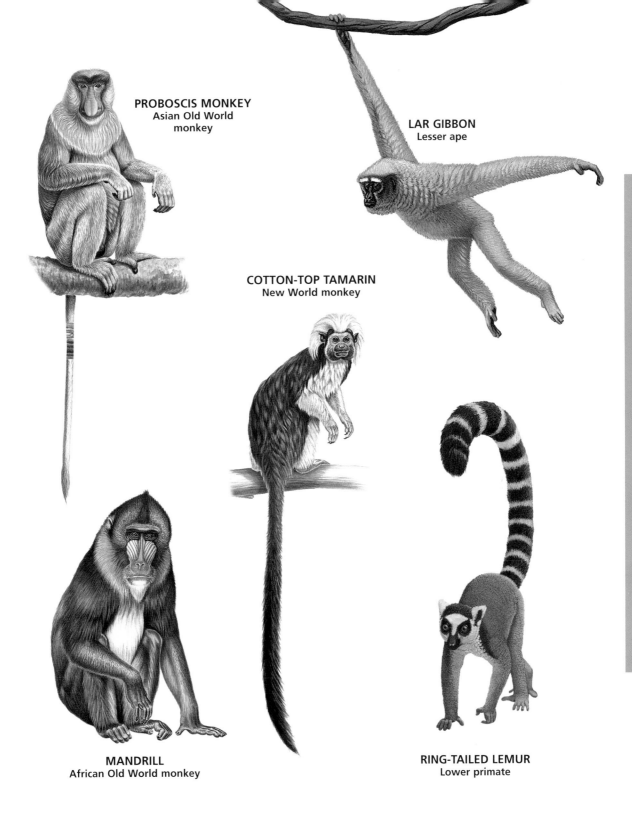

PROBOSCIS MONKEY
Asian Old World
monkey

LAR GIBBON
Lesser ape

COTTON-TOP TAMARIN
New World monkey

MANDRILL
African Old World monkey

RING-TAILED LEMUR
Lower primate

Lower Primates

Location

■ Lower primates

SENEGAL GALAGO
Galago senegalensis

VERREAUX'S SIFAKA
Propithecus verreauxi

INDRI
Indri indri

TARSIER
Tarsius sp.

GRAY MOUSE LEMUR
Microcebus murinus

Ring-tailed lemurs live in troops of up to 20 individuals.

Black and white rings on tail

Their diet consists of foraged fruit and insects.

LOWER PRIMATE ADAPTATIONS

An enlarged toilet claw, used for grooming, is only on the second digit.

Troop males use horny spurs on their wrists to mark their territory.

RING-TAILED LEMUR
Lemur catta

Dental comb on lower jaw

DENTAL COMB
Used in mutual grooming

Lower incisors and canines

Lower molar

Lower Primates

FORK-MARKED LEMUR
Phaner furcifer

GOLDEN POTTO
Arctocebus calabarensis

RED RUFFED LEMUR
Varecia variegata rubra

SPECTRAL TARSIER
Tarsius spectrum

DIADEMED SIFAKA
Propithecus diadema

BROWN MOUSE LEMUR
Microcebus rufus

COQUEREL'S MOUSE LEMUR
Microcebus coquereli

PHILIPPINE TARSIER
Tarsius syrichta

SLOW LORIS
Nycticebus coucang

AYE-AYE
Daubentonia madagascariensis

109

Higher Primates

NEW WORLD
Woolly monkey

OLD WORLD
Vervet monkey

Narrow septum
with nostrils that
face downward

Larger species
have a prehensile
tail that acts as
a fifth limb.

Thumbs on forefeet
and hindfeet are highly
opposable (able to
grasp objects easily).

Broad septum
with nostrils that
face sideways

Tail not prehensile

Thumbs are not
highly opposable
to other fingers.

SPIDER MONKEY

Spiders monkeys move through the trees using their extended limbs or swinging from their prehensile tail.

HIGHER PRIMATE FACES

Greater white-nosed monkey

Japanese macaque

Yellow baboon

Tonkin snub-nosed monkey

LAR GIBBON

Using their bodies as a pendulum, lar gibbons swing hand-over-hand in an energy-saving process called brachiation. They can travel up to 10 feet (3 m) each swing.

Ball-and-socket joints in the wrists increase mobility.

African Old World Monkeys

Location

■ African Old World monkeys

BLACK AND WHITE COLOBUS
Colobus guereza

ALLEN'S SWAMP MONKEY
Allenopithecus nigroviridis

AGILE MANGABEY
Cercocebus agilis

GELADA BABOON
Theropithecus gelada

DIANA MONKEY
Cercopithecus diana

HAMADRYAS BABOONS
Papio hamadryas

VERVET MONKEY
Chlorocebus aethiops

DRILL
Mandrillus leucophaeus

BLACK COLOBUS
Colobus satanas

113

Asian Old World Monkeys

Location

■ Asian Old World monkeys

DOUC LANGUR
Pygathrix nemaeus

PROBOSCIS MONKEY
Nasalis larvatus

**CELEBES CRESTED
MACAQUE**
Macaca nigra

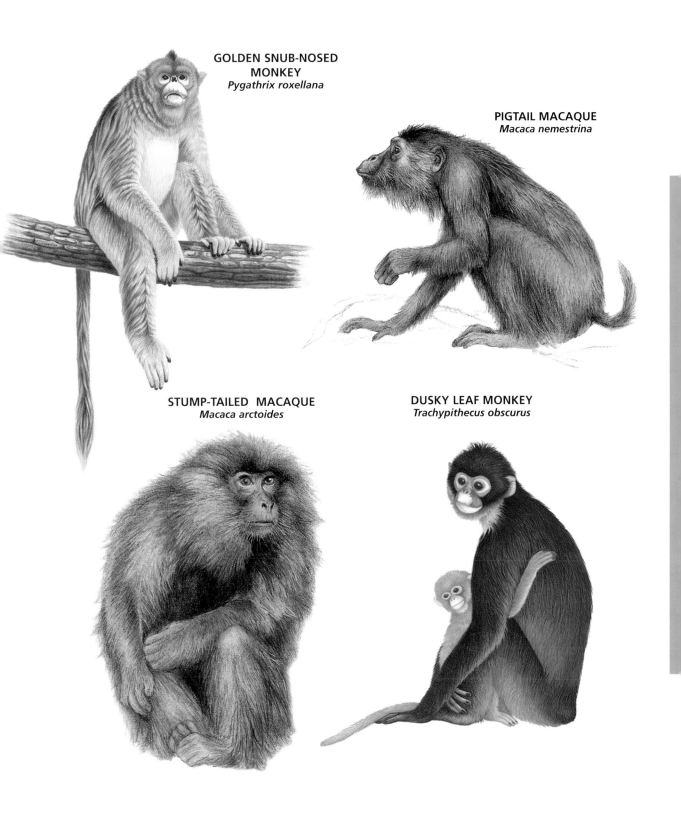

GOLDEN SNUB-NOSED MONKEY
Pygathrix roxellana

PIGTAIL MACAQUE
Macaca nemestrina

STUMP-TAILED MACAQUE
Macaca arctoides

DUSKY LEAF MONKEY
Trachypithecus obscurus

New World Monkeys

Location

■ New World monkeys

MIDAS TAMARIN
Saguinus midas

BROWN CAPUCHIN
Cebus apella

NORTHERN NIGHT MONKEY
Aotus trivirgatus

RED HOWLER MONKEY
Alouatta seniculus

MURIQUI
Brachyteles arachnoides

COTTON-TOP TAMARINS
Saguinus oedipus
Cotton-top tamarins live in family troops of 10–12 individuals with a single breeding pair. Older siblings and troop males help raise and carry the pair's offspring.

RED UAKARI
Cacajao calvus

MASKED TITI
Callicebus personatus

Great Apes

Location

■ Apes

MOUNTAIN GORILLA
Gorilla gorilla beringei

WESTERN LOWLAND GORILLA
Gorilla gorilla gorilla

Head of male orangutan

ORANGUTAN
Pongo pygmaeus

CHIMPANZEE
Pan troglodytes

BONOBO
Pan paniscus

119

Gorillas

SKULL COMPARISON

Lower primate

Lemur skull

Higher primate

Gorilla skull

Human skull

FAMILY GROUPS

Mountain gorillas live in closely knit family groups of related females in the mountain forests of Africa.

A large and dominant silverback male leads and protects the group.

Gorillas build nests in the trees each night for warmth and safety.

All apes have a fully opposable thumb that is used to grasp objects.

Gorillas are plant eaters (herbivores) and spend most of the day eating.

Carnivores

EVOLUTION OF CARNIVORES

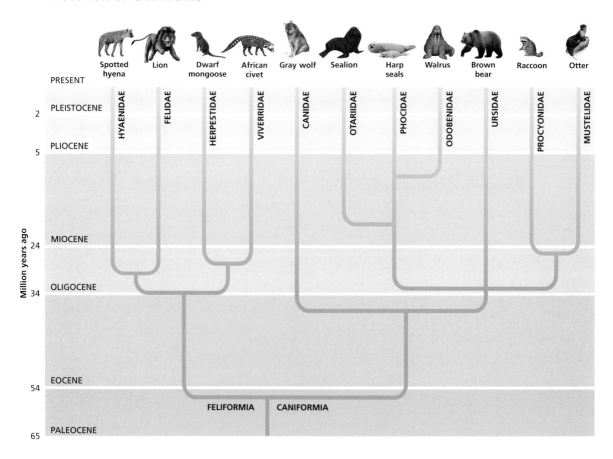

Spotted hyena · Lion · Dwarf mongoose · African civet · Gray wolf · Sealion · Harp seals · Walrus · Brown bear · Raccoon · Otter

PRESENT · PLEISTOCENE · PLIOCENE · MIOCENE · OLIGOCENE · EOCENE · PALEOCENE

2 · 5 · 24 · 34 · 54 · 65

Million years ago

HYAENIDAE · FELIDAE · HERPESTIDAE · VIVERRIDAE · CANIDAE · OTARIIDAE · PHOCIDAE · ODOBENIDAE · URSIDAE · PROCYONIDAE · MUSTELIDAE

FELIFORMIA · CANIFORMIA

Carnivores

MAMMALS

DENTAL ADAPTATIONS

■ Slicing teeth ■ Grinding teeth

Cat: highly carnivorous

Dog: omnivorous

Brown bear: omnivorous

Giant panda: herbivorous

TYPES OF CARNIVORES

Ocelot
Family *Felidae*

Pharoah hound
Family *Canidae*

Brown bear
Family *Ursidae*

Raccoon
Family *Procyonidae*

Weasel
Family *Mustelidae*

African civet
Family *Viverridae*

Spotted hyena
Family *Hyaenidae*

Pounces
silently

Grabs prey
by throat

Prey unable
to breathe

Tiger kills prey
by suffocation.

Tiger tracks prey
while camouflaged
by long grass.

TIGER HUNTING

Wolves, Dogs and Foxes

GRAY WOLF PACK

Wolves are efficient hunters who work in a group to exhaust their prey.

Wolf cubs stay with their birth pack until they mature physically.

Wolves communicate with howls, body language and facial expressions.

Wolf packs enforce a strict social hierarchy within the group.

Gray wolf

Female red fox with pups

African wild dogs fighting

WOLF TAILS

Threatening

Relaxed

Fearful

Submissive

Ready to attack

DOMESTIC DOG

DOMESTIC DOG SKELETON

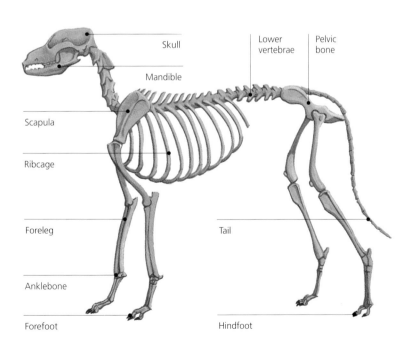

Skull

Mandible

Scapula

Ribcage

Foreleg

Anklebone

Forefoot

Lower vertebrae

Pelvic bone

Tail

Hindfoot

Wolves, Dogs and Foxes

Location

- Wolves and wild dogs
- Foxes
- Wolves, wild dogs and foxes

RACCOON DOG
Nyctereutes procyonoides

THREE TYPES OF DOMESTIC DOGS

Chihuahua
Toy dog

Fox terrier
Hunting dog

Samoyed
Working dog

Dalmatian

Rescue dog

Dog sled

Chihuahua

Sheepdog

PRESENT

Thousand years ago

5

10

15

DOG TRAVOIS

ANCESTRAL WOLF

MANED WOLF
Chrysocyon brachyurus

GRAY WOLF
Canis lupus

AFRICAN WILD DOG
Lycaon pictus

BUSH DOG
Speothos venaticus

DINGO
Canis lupus dingo

RED WOLF
Canis rufus

Wolves, Dogs and Foxes

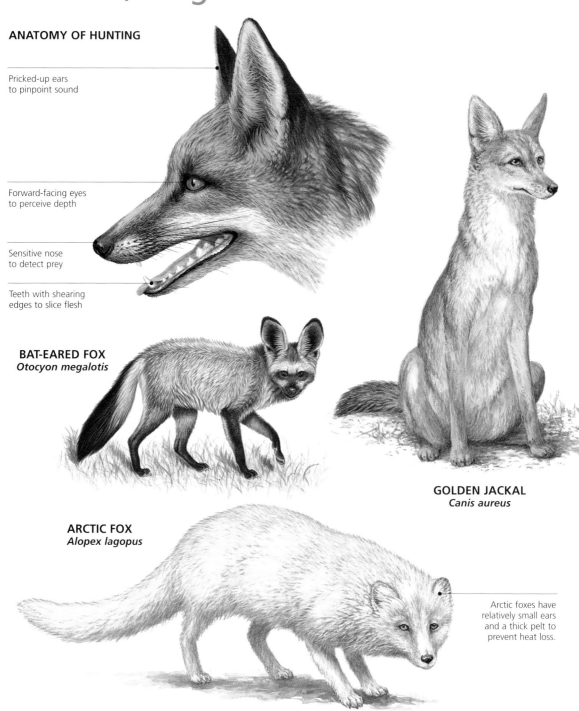

ANATOMY OF HUNTING

Pricked-up ears
to pinpoint sound

Forward-facing eyes
to perceive depth

Sensitive nose
to detect prey

Teeth with shearing
edges to slice flesh

BAT-EARED FOX
Otocyon megalotis

GOLDEN JACKAL
Canis aureus

ARCTIC FOX
Alopex lagopus

Arctic foxes have
relatively small ears
and a thick pelt to
prevent heat loss.

COYOTE
Canis latrans

GRAY FOX
Urocyon cinereoargenteus

KIT FOX
Vulpes macrotis

COYOTE EXPRESSIONS

Submissive

Combative

Defensive

Friendly

Playful

Bears

EVOLUTION OF BEARS

PRESENT | Red panda | Raccoon | Giant panda | Spectacled bear | Sloth bear | Sun bear | American black bear | Asiatic black bear | Polar bear | Brown bear

Million years ago

2 PLEISTOCENE

5 PLIOCENE

24 MIOCENE

34 OLIGOCENE

AILURINAE PROCYONINAE AILUROPODINAE TREMARCTINAE MELURSUS HELARCTOS URSINAE

PROCYONIDAE URSIDAE

BEAR SIZES

Sun bear

Giant panda

Asiatic black bear

Sloth bear

American black bear

Spectacled bear

Brown bear

Polar bear

FOOT COMPARISON

American black bear
Long, sharp claws on forefeet

Giant panda
Short, blunt claws on forefeet

Walks flat on hindfeet

Walks on toes of hindfeet

FOREFOOT COMPARISON

Brown bear
Long claws for digging

Giant panda
Modified wrist acts like digit

BROWN BEAR

TREE CLIMBERS
American black bears are agile climbers and often snooze in the safety of the treetops.

BROWN BEAR SKELETON

Skull

Mandible

Scapula

Ribcage

Pelvic bone

Hip

Knee

Foreleg

Anklebone

Claws

131

Eurasian and New World Bears

Location

- Eurasian bear (brown bear)
- New World bears
- Eurasian bear and New World bears

BROWN BEAR
Ursus arctos

AMERICAN BLACK BEAR REPRODUCTION

Winter
Hibernation and birth of young

Spring
Nursing and rearing young

Summer
Mating and intense feeding

Autumn
Delayed implantation of embryo

SPECTACLED BEAR
Tremarctos ornatus

**AMERICAN
BLACK BEAR**
Ursus americanus

Polar Bear

Location

■ Polar bear

POLAR BEAR CUBS

POLAR BEAR YEAR

1. Late February–April
Mother and pups leave den and mother replenishes fat stores.

2. March–April
Adult males travel in search of breeding females, then fight for breeding rights.

8. December–February
Except for pregnant females, polar bears actively feed throughout winter.

7. November–January
Pregnant female gives birth and nurses cubs in the den.

POLAR BEAR
Ursus maritimus

3. April–May
Mating occurs over two weeks and eggs are fertilized, but are not implanted until September.

4. May–June
At two-and-a-half years of age, young bears are weaned from their mothers.

6. August–late November
Sea ice melts and polar bears come ashore, living off their fat reserves until the sea freezes again.

5. April–July
Ringed seal pups, born in April, are easy prey for polar bears.

135

Asian Bears

Location

- Asian bears
- Giant panda

SLOTH BEAR
Melursus ursinus

ASIATIC BLACK BEAR
Ursus thibetanus

SUN BEAR
Reaching only 3 feet (1 m) in height, the sun bear is the world's smallest.

SUN BEAR
Helarctos malayanus

GIANT PANDA
Ailuropoda melanoleuca

137

Mustelids

Location

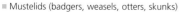

■ Mustelids (badgers, weasels, otters, skunks)

AMERICAN BADGER
Taxidea taxus

EUROPEAN BADGER SETT (BURROW)
While badgers are usually solitary creatures, they build an elaborate system of communal tunnels, called a sett, with individual nesting chambers for each badger.

Entrance

Individual chamber

Individual chamber

EUROPEAN POLECAT
Mustela putorius

MARBLED POLECAT
Vormela peregusna

EURASIAN BADGER
Meles meles

HOG BADGER
Arctonyx collaris

HUMBOLDT'S HOG-NOSED SKUNK
Conepatus humboldtii

MOLINA'S HOG-NOSED SKUNK
Conepatus chinga

SKUNK DEFENSE
Mustelids have musk glands
that are used for defense
and marking territory.

Confronts predator face-to-face

Raises tail in warning

Rises up on forelegs

Ejects foul-smelling liquid

Mustelids

SEA OTTER
Sea otters float on their back and use small stones to crack open sea urchins.

MARTEN
Martens are agile hunters with powerful forelimbs and a long tail. They chase prey by leaping through the trees.

EUROPEAN OTTER
Lutra lutra

GIANT OTTER
Pteronura brasiliensis

ERMINE
Mustela erminea
Summer coat

AFRICAN STRIPED WEASEL
Poecilogale albinucha

ERMINE
Winter coat

LEAST WEASEL
Mustela nivalis

EUROPEAN MINK
Mustela lutreola

YELLOW-THROATED MARTEN
Martes flavigula

EUROPEAN PINE MARTEN
Martes martes

Catlike Carnivores

Location

■ Catlike carnivores

Cobra rises
up to strike.

MONGOOSE AND COBRA
Some mongooses prey on venomous
cobras. While not immune to cobras'
venom, mongooses are fast, agile
hunters who grab their prey behind
the head and avoid their strike.

ASIAN PALM CIVET
Paradoxurus hermaphroditus

MEERKATS
Suricata suricatta

Meerkats live in family groups, complete with sentinels to protect the group from predators.

FOSSA
Cryptoprocta ferox

BANDED LINSANG
Prionodon linsang

AFRICAN CIVET
Civettictis civetta

Raccoons and Hyenas

Location

■ Raccoons and relatives
■ Hyenas and aardwolf

KINKAJOU
Potos flavus

SPOTTED HYENA
Crocuta crocuta

STRIPED HYENA
Hyaena hyaena

RED PANDA
Ailurus fulgens

SOUTH AMERICAN COATI
Nausa nausa

AARDWOLF
Proteles cristatus

RINGTAIL
Bassariscus astutus

NORTHERN RACCOON
Procyon lotor

Hyena Hunting Tactics

PACK HUNTING

The largest members of the hyena family, spotted hyenas are not exclusively scavengers, but are efficient pack hunters.

They are able to accelerate to speeds of over 30 miles per hour (48 km/h), and will chase their prey at this speed for more than a mile.

Other hyenas nearby join in the chase, until the gemsbok is exhausted. They work together to bring down their prey.

The spotted hyena's jaws are strong enough to crunch through bone; the hydrochloric acid in its stomach breaks down bone.

Rapid eaters, a pack of hyenas can demolish an adult gemsbok in as little as half an hour, this speed necessitated by threats from scavengers.

Despite their reputation as fearsome hunters, lions are just as likely to scavenge a kill from hyenas, who are then forced to abandon it.

STEALING THE KILL

Cats

EVOLUTION OF CATS

MARGAY EXPRESSIONS

Aggressive

Defensive

SPRINGLIKE CLAWS

Muscles and tendons relaxed

Contracted to expose claw

WILD CAT

WILD CAT SKELETON

Skull

Enlarged
neck vertebra

Scapula

Sternum

Ribcage

Lower
vertebrae

Tail

Hyoid (supports
larynx and
base of tongue)

Pelvic bone

Humerus

Femur

Patella
(kneebone)

Radius

Ulna

Tibia

Carpals

Phalanges

Metacarpals

Metatarsals
(toe bones)

Great Cats

Location

■ Great cats

OCELOT
Leopardus pardalis

LION PRIDE

There are usually one, but often two related dominant males in the pride.

Lionesses are the primary hunters of the pride, and hunt together to provide meat for the group.

Females breed from the age of four, and mate every 20 minutes for five days when on heat.

Up to three generations of females can live in one pride, which can hold as many as 30 individuals.

Lion cubs nurse for up to six months, before weaning. Cubs are the last to feed from kills.

PUMA
Puma concolor

LION AND LIONESS
Panthera leo

TIGER
Panthera tigris

Great Cats

CLOUDED LEOPARD
Neofelis nebulosa

MARBLED CAT
Pardofelis marmorata

SNOW LEOPARD
Uncia uncia

LEOPARD AND GIRAFFE
Leopards often rest in trees, and will stow a kill on a branch to keep it out of reach of competitors and scavengers.

JAGUAR
Panthera onca

BLACK PANTHER
Panthera pardus
Also known as melanic leopards, panthers have a genetic mutation that raises the amount of melanin (pigment) in the skin and fur.

LEOPARD
Panthera pardus

Lynxes and Cheetah

Location

- Lynxes, caracal and bobcat
- Lynxes, caracal, bobcat and cheetah

EURASIAN LYNX
Lynx lynx

CANADA LYNX
Lynx canadensis

BOBCAT
Lynx rufous

CARACAL
Caracal caracal

SPANISH LYNX
Lynx pardinus

CHEETAH
Acinonyx jubatus

FASTEST ON LAND
Cheetahs can run at speeds of 70 miles
(110 km) per hour in short bursts of speed.

Small Cats

Location

■ Small cats

KODKOD
Oncifelis guigna

LEOPARD CAT
Prionailurus bengalensis

JUNGLE CAT
Felis chaus

FISHING CAT
Prionailurus viverrinus

ANDEAN MOUNTAIN CAT
Felis jacobita

SERVAL
Leptailurus serval

ASIAN GOLDEN CAT
Profelis temmincki

AFRICAN GOLDEN CAT
Profelis aurata

JAGUARUNDI
Herpailurus yagouaroundi

MARGAY
Felis wiedii

157

Seals, Sealions and Walrus

FUR SEAL

Layer of blubber and thick fur to protect from cold

SEALION SKULL

Parallel rows of teeth

Flattened auditory bullae

Nasal cavity

Eye socket

Sagittal crest

AUSTRALIAN SEALION COLONY

Australian sealions live in loosely structured colonies on the coast.

Males fight for, and then defend, breeding territory in the colony.

Sealion pups are agile swimmers from a few months of age.

Sealions feed on fish, shellfish and octopus, often hunting together.

WALRUS

WALRUS SKELETON

Tusks can grow to 3 feet (1 m).

Skull

Neck vertebrae

Scapula

Ribcage

Lower vertebrae

Hindleg propels walrus through the water.

Pelvic bone

Foreleg is used like a rudder.

Tail

Forefoot (flipper)

Hindfoot (flipper)

Seals, Sealions and Walrus

Location
- Seals and sealions
- Seals, sealions and walrus

BEARDED SEAL
Erignathus barbatus

HARP SEAL
Phoca groenlandica

STELLER SEALION
Eumetopias jubatus

NEW ZEALAND FUR SEAL
Arctocephalus forsteri

RIBBON SEAL
Phoca fasciata

BAIKAL SEAL
Phoca sibirica

Harbor seal with young

Bull elephant seal guarding territory

Leopard seal

Walrus tusks are used for fighting.

Fur seal showing hind flippers

Hooded seal with inflatable skin pouch

Harp seal pups

Crabeater seals do not eat crabs.

Weddell seals diving off pack ice

Whales, Dolphins and Porpoises

EVOLUTION OF WHALES

Yangtze river dolphin

Narwhal

Common dolphin

Strap-toothed whale

Pygmy sperm whale

Pygmy right whale

Blue whale

Amazon river dolphin

Dall's porpoise

Ganges river dolphin

Sperm whale

Bowhead whale

Gray whale

Hippopotamus

Million years ago

PRESENT

PLEISTOCENE

2

PLIOCENE

5

MIOCENE

24

OLIGOCENE

34

EOCENE

54

PONTOPORIIDAE

INIIDAE

MONODONTIDAE

PHOCOENIDAE

DELPHINIDAE

PLATANISTIDAE

ZIPHIIDAE

PHYSETERIDAE

KOGIIDAE

BALAENIDAE

NEOBALAENIDAE

ESCHRICHTIIDAE

BALAENOPTERIDAE

ARTIODACTLYS

ODONTOCETI

MYSTICETI

CETACEA

Gray whale

TYPES OF WHALES
The sei whale and the gray whale are both baleen whales, while the killer whale is a dolphin, which is a type of toothed whale.

Sei whale

Killer whale

WHALE TAILS

Common dolphin

Southern right whale

Blue whale

Humpback whale

Gray whale

Killer whale

Whale Anatomy

Sperm whale head

Bottlenose dolphin head

TOOTHED WHALE
Age is measured by counting rings inside the teeth.

TOOTHED WHALE
Atlantic white-sided dolphin

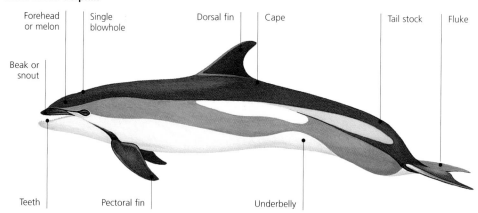

Forehead or melon

Single blowhole

Dorsal fin

Cape

Tail stock

Fluke

Beak or snout

Teeth

Pectoral fin

Underbelly

TOOTHED WHALE SKELETON

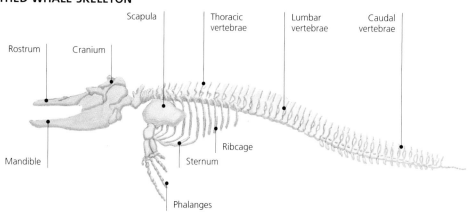

Scapula

Thoracic vertebrae

Lumbar vertebrae

Caudal vertebrae

Rostrum

Cranium

Mandible

Sternum

Ribcage

Phalanges

Minke whale head

Southern right whale head

BALEEN WHALE

Age is measured by counting rings inside the waxy ear plug.

BALEEN WHALE
Pygmy right whale

Baleen

Double
blowhole

Dorsal fin

Keel

Tail stock

Fluke

Rostrum

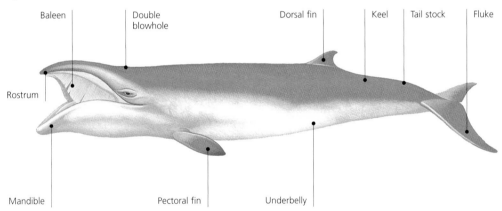

Mandible

Pectoral fin

Underbelly

BALEEN WHALE SKELETON

Rostrum

Thoracic
vertebrae

Lumbar
vertebrae

Caudal
vertebrae

Cranium

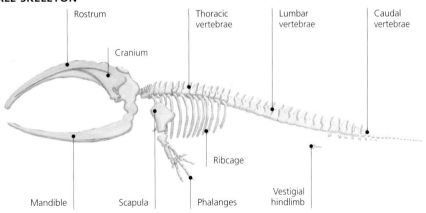

Ribcage

Mandible

Scapula

Phalanges

Vestigial
hindlimb

Baleen Whales

BUBBLENETTING

Fish become trapped and move closer together.

Humpback whale feeds easily off the trapped fish.

Humpback whale blows bubbles around school of fish.

Both whales swim in a spiral around the school of fish.

DIVING FOR FOOD

Krill

Fin whale inhales.

It rolls into steep dive.

BALEEN WHALE BEHAVIOR

Lobtailing

Pecslapping

Spyhopping

Breaching

MINKE WHALE

Although many minke whales head to warmer waters for winter breeding, others stay behind. These whales are usually not ready to breed, so save their energy for the following year.

BALEEN WHALE SPOUTS

Humpback whale

Sei whale

Right whale

The whale uses hearing to find prey.

Swarm of krill or zooplankton

It filters food from mouthfuls of water.

The fin whale surfaces.

Baleen Whales

Location

■ Baleen whales

PYGMY RIGHT WHALE
Caperea marginata

GRAY WHALE WITH CALF
Eschrichtius robustus

FIN WHALE
Balaenoptera physalus

MINKE WHALE
Balaenoptera acutorostrata

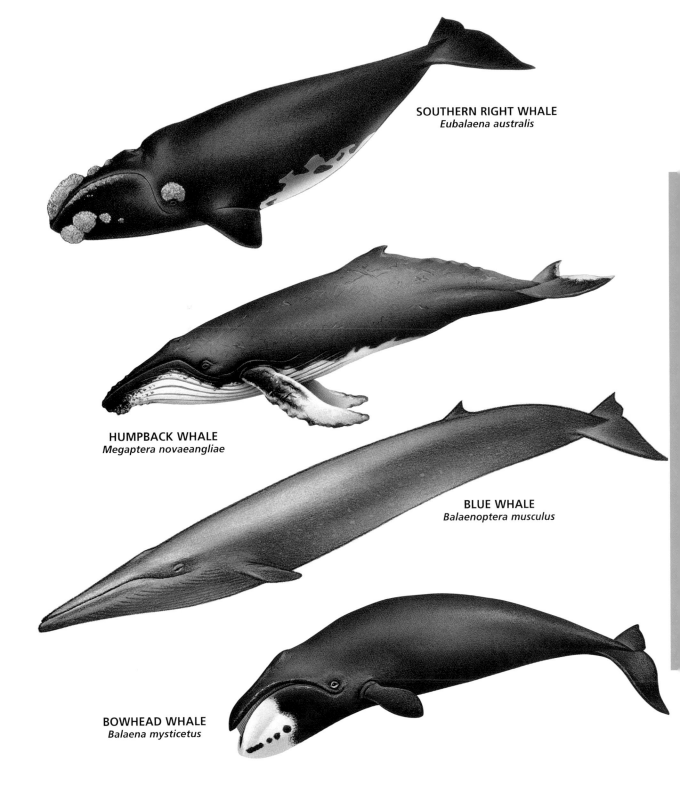

SOUTHERN RIGHT WHALE
Eubalaena australis

HUMPBACK WHALE
Megaptera novaeangliae

BLUE WHALE
Balaenoptera musculus

BOWHEAD WHALE
Balaena mysticetus

Toothed Whales

DIVE DEPTHS

Some whales are able to dive to deeper depths than others in search of food. Their bodies are adapted to the intense pressure and cold of deep waters and they can hold their breath for longer periods of time.

Franciscana: 30 feet (9 m)

Short-beaked common dolphin: 330 feet (100 m)

Short-finned pilot whale: 990 feet (300 m)

Sperm whale: 10,000 feet (3,000 m)

MAKING MUSIC

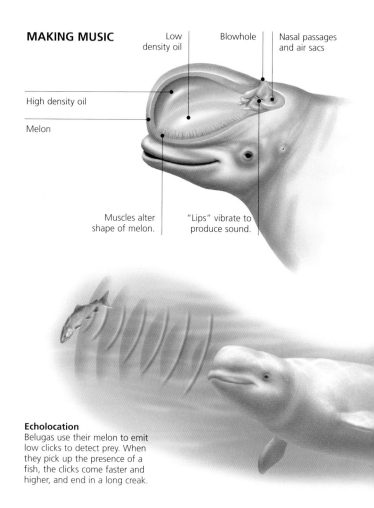

Low density oil

Blowhole

Nasal passages and air sacs

High density oil

Melon

Muscles alter shape of melon.

"Lips" vibrate to produce sound.

Echolocation

Belugas use their melon to emit low clicks to detect prey. When they pick up the presence of a fish, the clicks come faster and higher, and end in a long creak.

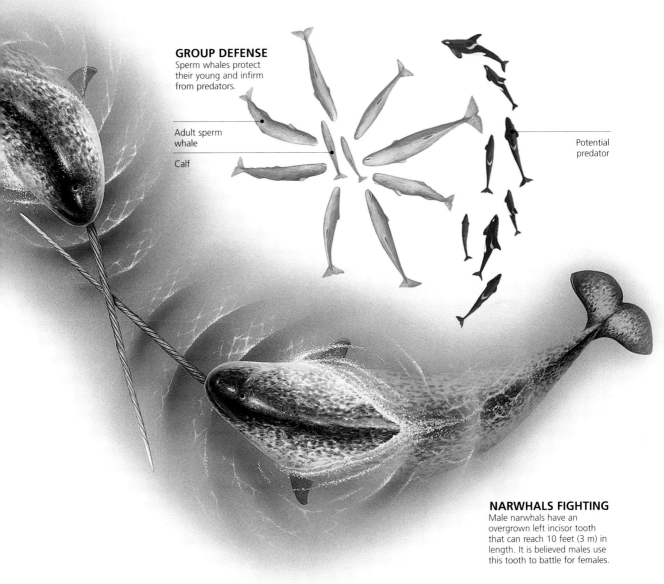

GROUP DEFENSE
Sperm whales protect their young and infirm from predators.

Adult sperm whale

Calf

Potential predator

NARWHALS FIGHTING
Male narwhals have an overgrown left incisor tooth that can reach 10 feet (3 m) in length. It is believed males use this tooth to battle for females.

Types of teeth

Narwhal

Ganges river dolphin

Sperm whale

Strap-toothed whale

Harbor porpoise

Toothed whale behavior

Touching beaks

Touching pectoral fins

Toothed Whales

Location

■ Toothed whales

STRAP-TOOTHED WHALE
Mesoplodon layardii

SPERM WHALE DIVING | Sperm whale inhales and prepares to dive. | It dives vertically to about 1,300 feet (394 m), but some may descend to 10,000 feet (3,000 m). | It hunts in almost total darkness for prey such as jumbo flying squid and giant squid. | Giant squid

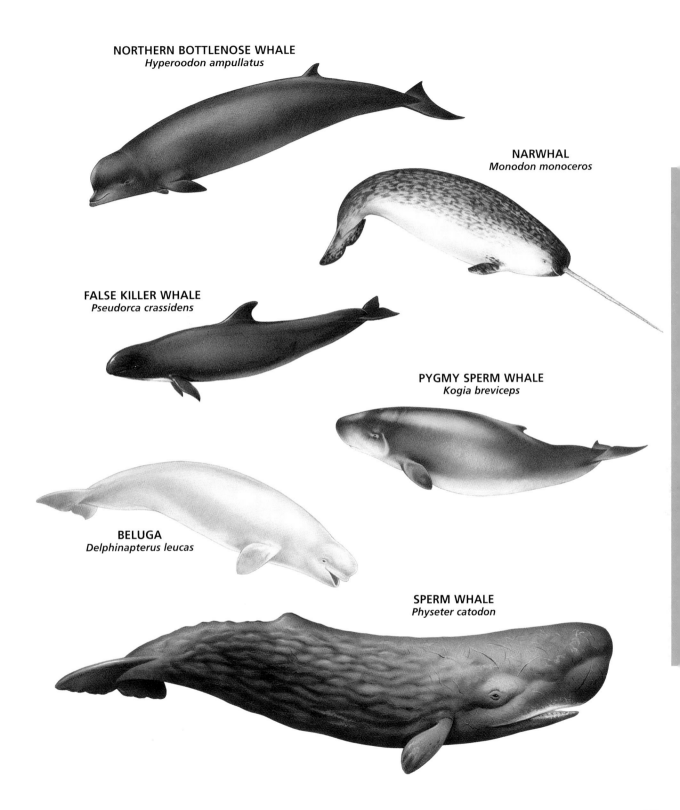

NORTHERN BOTTLENOSE WHALE
Hyperoodon ampullatus

NARWHAL
Monodon monoceros

FALSE KILLER WHALE
Pseudorca crassidens

PYGMY SPERM WHALE
Kogia breviceps

BELUGA
Delphinapterus leucas

SPERM WHALE
Physeter catodon

Dolphins

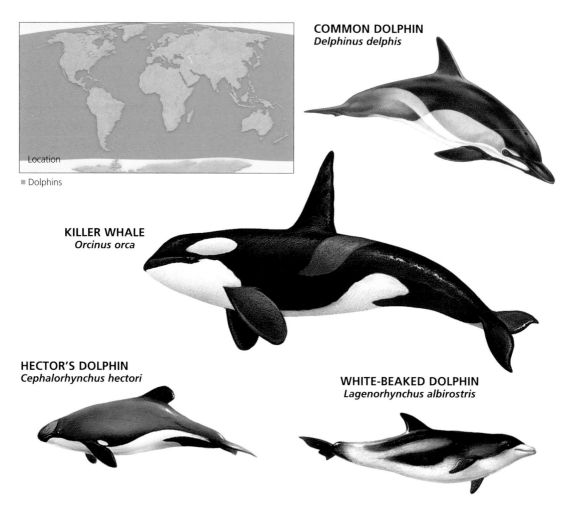

Location

■ Dolphins

COMMON DOLPHIN
Delphinus delphis

KILLER WHALE
Orcinus orca

HECTOR'S DOLPHIN
Cephalorhynchus hectori

WHITE-BEAKED DOLPHIN
Lagenorhynchus albirostris

DOLPHIN PROPULSION

| Raised flukes | Pushes down with flukes | Moves forward and upward | Pushes up with flukes | Moves forward and downward |

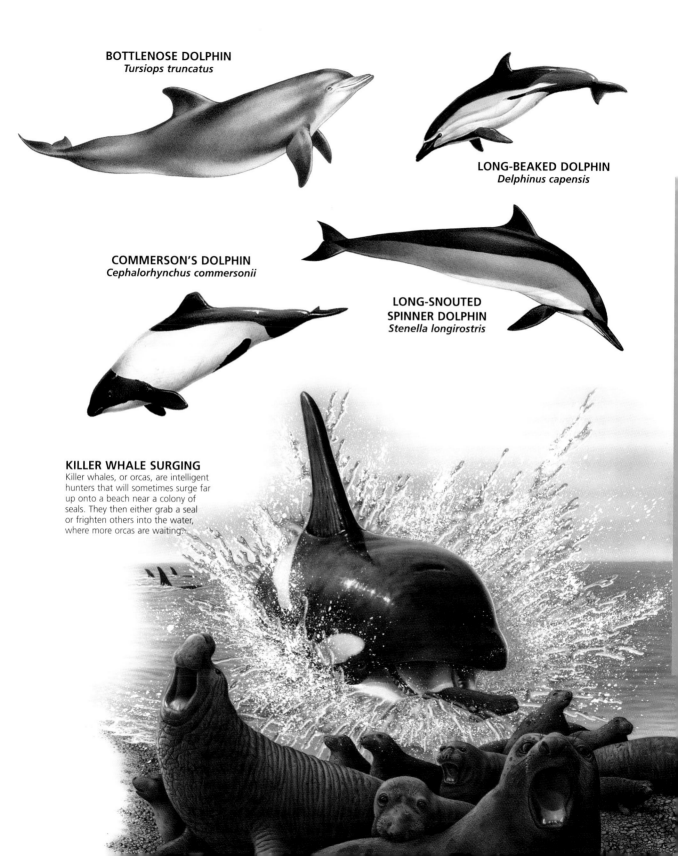

BOTTLENOSE DOLPHIN
Tursiops truncatus

LONG-BEAKED DOLPHIN
Delphinus capensis

COMMERSON'S DOLPHIN
Cephalorhynchus commersonii

**LONG-SNOUTED
SPINNER DOLPHIN**
Stenella longirostris

KILLER WHALE SURGING
Killer whales, or orcas, are intelligent
hunters that will sometimes surge far
up onto a beach near a colony of
seals. They then either grab a seal
or frighten others into the water,
where more orcas are waiting.

River Dolphins and Porpoises

Location

- River dolphins
- Porpoises
- River dolphins and porpoises

RIVER DOLPHIN ANATOMY

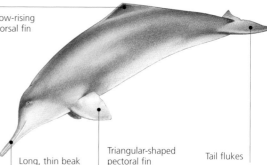

Low-rising dorsal fin

Long, thin beak

Triangular-shaped pectoral fin

Tail flukes

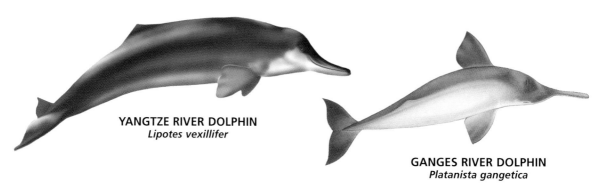

YANGTZE RIVER DOLPHIN
Lipotes vexillifer

GANGES RIVER DOLPHIN
Platanista gangetica

INDUS RIVER DOLPHIN
Platanista minor

AMAZON RIVER DOLPHIN
Inia geoffrensis

VAQUITA
Phocoena sinus

HARBOR PORPOISE
Phocoena phocoena

SPECTACLED PORPOISE
Australophocaena dioptrica

FINLESS PORPOISE
Neophocaena phocaenoides

DALL'S PORPOISE
Phocoenoides dalli

Hoofed Mammals

EVOLUTION OF HOOFED MAMMALS

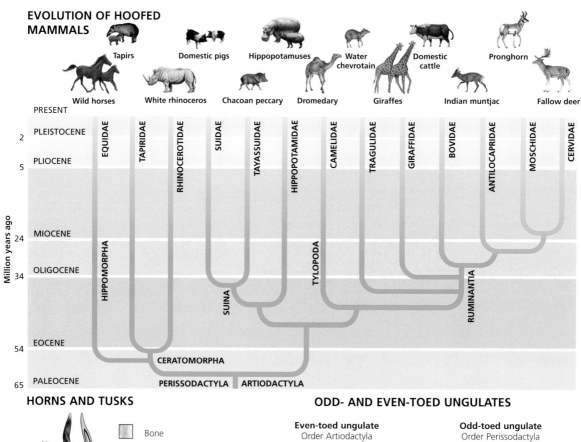

Tapirs
Domestic pigs
Hippopotamuses
Water chevrotain
Domestic cattle
Pronghorn

Wild horses
White rhinoceros
Chacoan peccary
Dromedary
Giraffes
Indian muntjac
Fallow deer

Million years ago

PRESENT

PLEISTOCENE — 2

PLIOCENE — 5

MIOCENE — 24

OLIGOCENE — 34

EOCENE — 54

PALEOCENE — 65

EQUIDAE · TAPIRIDAE · RHINOCEROTIDAE · SUIDAE · TAYASSUIDAE · HIPPOPOTAMIDAE · CAMELIDAE · TRAGULIDAE · GIRAFFIDAE · BOVIDAE · ANTILOCAPRIDAE · MOSCHIDAE · CERVIDAE

HIPPOMORPHA · SUINA · TYLOPODA · RUMINANTIA

CERATOMORPHA

PERISSODACTYLA · ARTIODACTYLA

HORNS AND TUSKS

☐ Bone
■ Bone that is shed
☐ Keratin

ODD- AND EVEN-TOED UNGULATES

Even-toed ungulate
Order Artiodactyla

Odd-toed ungulate
Order Perissodactyla

Carpals

Carpals

Metacarpals

Metacarpals

Phalanges

Phalanges

Hoof (toenail)

Hoof (toenail)

ZEBRA
Odd-toed ungulate

DROMEDARY
Even-toed ungulate

Walks on central toe of each foot. Other toes are vestigial.

Walks on third and fourth toes of each foot. No other toes remain.

WHITE RHINOCEROS
Odd-toed ungulate

REINDEER
Even-toed ungulate

Walks on middle three toes of front foot. No other toes remain.

Walks on four toes of each foot. No other toes remain.

Hippopotamuses, Pigs and Peccaries

Location

- Wild pigs
- Peccaries
- Hippopotamuses and wild pigs

HIPPOPOTAMUS HEAD

Fleshy lips used to pull out grass

Tusklike lower canine used for territorial fighting

Lower incisor

Eyes, ears and nostrils located on top of head

HIPPOPOTAMUS
Hippopotamus amphibius

PYGMY HIPPOPOTAMUS
Hexaprotodon liberiensis

CHACOAN PECCARY
Catagonus wagneri

DOMESTIC PIG
Sus scrofa

BABIRUSA
Babyrousa babyrussa

RED RIVER HOG
Potamochoerus porcus

BEARDED PIG
Sus barbatus

WARTHOG
Phacochoerus africanus

Old World Camels

Location

■ Bactrian camels (wild dromedary camels extinct)

CAMEL HUMP

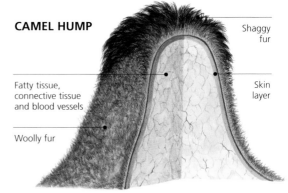

Shaggy fur

Skin layer

Fatty tissue, connective tissue and blood vessels

Woolly fur

CAMEL ANATOMY

Horny padding on hindlegs protects camels from heat.

Nostrils can be closed in sandstorms.

Camels walk two legs at a time, right following left, which causes their body to rock and sway.

Two rows of eyelashes keep out sand.

Thick fatty pads stop feet sinking.

Twisted nasal passages reabsorb water.

DROMEDARY
Camelus dromedarius

BACTRIAN CAMEL
Camelus bactrianus

Arabian camel-racing

Algerian nomad with camel

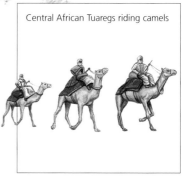

Central African Tuaregs riding camels

New World Camelids

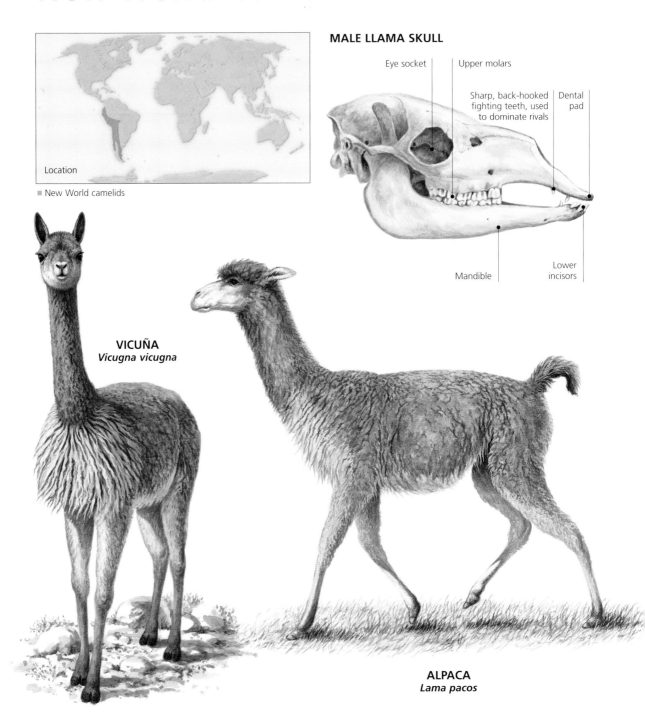

Location

■ New World camelids

MALE LLAMA SKULL

Eye socket

Upper molars

Sharp, back-hooked
fighting teeth, used
to dominate rivals

Dental
pad

Mandible

Lower
incisors

VICUÑA
Vicugna vicugna

ALPACA
Lama pacos

LLAMA
Lama glama

GUANACO PREPARING TO SPIT
Lama guanicoe

Domestic alpaca ready for shearing

Guancos alert for predators

Domestic llamas bred as pack animals

Giraffe, Okapi and Pronghorn

Location

■ Giraffe and okapi
■ Pronghorn

GIRAFFE
Giraffa camelopardalis

FEEDING ON ACACIA

Prehensile
tongue can be
over 12 inches
(30 cm) long.

The giraffe grabs the
leaves with its tongue
and rips off branches
with its teeth.

Acacia
thorns

GIRAFFE DRINKING
Because they have such
long legs and neck,
giraffes must splay their
front legs wide to bend
low enough to drink.

PRONGHORN TAKING FLIGHT
The fastest animal in the New World, the pronghorn flashes its white underside in warning as it bounds away from danger.

PRONGHORN
Antilocapra americana

OKAPI
Okapia johnstoni

Deer

SPREADING MOOSE ANTLERS

One year
Horns are simple spikes.

Four years
Rack expands in size
and number of points.

Eight years
Mature rack demonstrates moose's
suitability as a potential mate.

MALE CARIBOU FIGHTING

Male caribou fight by
locking antlers during
the breeding season.

Fighting males rarely
become entangled by
their racks of antlers.

After pushing to see which is
stonger, the victor wins the right
to mate with the herd's females.

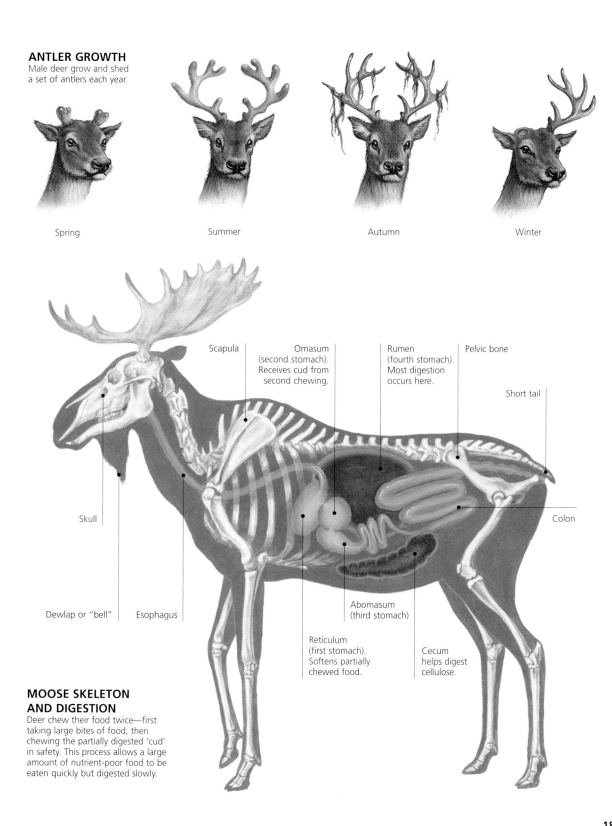

ANTLER GROWTH
Male deer grow and shed a set of antlers each year.

Spring

Summer

Autumn

Winter

Scapula

Omasum (second stomach). Receives cud from second chewing.

Rumen (fourth stomach). Most digestion occurs here.

Pelvic bone

Short tail

Skull

Colon

Dewlap or "bell"

Esophagus

Abomasum (third stomach)

Reticulum (first stomach). Softens partially chewed food.

Cecum helps digest cellulose.

MOOSE SKELETON AND DIGESTION
Deer chew their food twice—first taking large bites of food, then chewing the partially digested 'cud' in safety. This process allows a large amount of nutrient-poor food to be eaten quickly but digested slowly.

189

Deer

Location

■ Deer, musk deer and chevrotains

PUDU
Pudu sp.

WATER CHEVROTAIN
Hyemoschus aquaticus

BARASINGHA
Cervus duvaucelii

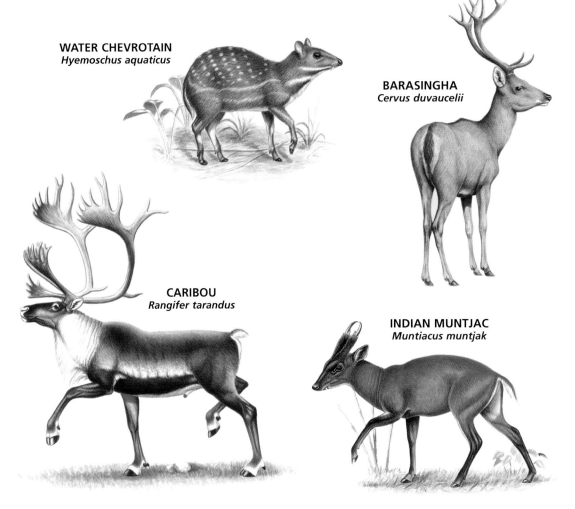

CARIBOU
Rangifer tarandus

INDIAN MUNTJAC
Muntiacus muntjak

FALLOW DEER
Dama dama

MOOSE
Alces alces

WAPITI
Cervus elaphus canadensis

Bovids

GERENUKS EATING ACACIA

Acacia shrubs

Narrow mouth
and face help the
gerenuk get close
to the leaves.

Forelegs help
the gerenuk to
balance upright.

Long, slender neck

The spine curves
to allow the
gerenuk to stand
up on hindlegs.

Weight is distributed
evenly onto the
hindlegs and hooves.

MOUNTAIN SHEEP FIGHTING

Rams approach one another, horns first.   They kick and ram each other, aiming for chest and belly.   They rear up on their hindlegs.

TYPES OF BOVIDS

Barbary sheep
Goat

Gaur
True cattle

Greater kudu
Antelope

Then drop with great force into a head-on clash.

They then present to each other.

The rams break from fighting to feed.

Antelopes

Location

■ Antelopes, gazelles and duikers

SABLE ANTELOPE
Hippotragus niger

DUIKER
Cephalophus sp.

KLIPSPRINGER
Oreotragus oreotragus

BONGO
Tragelaphus eurycerus

BLUE WILDEBEEST
Connochaetus taurinus

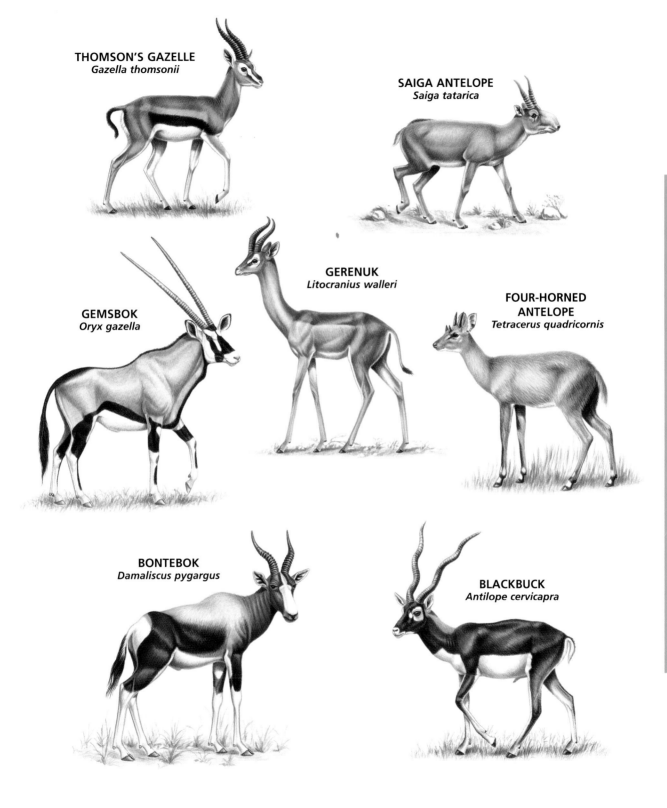

THOMSON'S GAZELLE
Gazella thomsonii

SAIGA ANTELOPE
Saiga tatarica

GERENUK
Litocranius walleri

FOUR-HORNED ANTELOPE
Tetracerus quadricornis

GEMSBOK
Oryx gazella

BONTEBOK
Damaliscus pygargus

BLACKBUCK
Antilope cervicapra

Wildebeest Migration

Location

■ Wildebeest

Migratory path

■ Serengeti National Park
■ Migration through Serengeti-Mara ecosystem

ACROSS THE SERENGETI

Each year, thousands of wildebeest make their way across the Serengeti–Mara ecosystem in Tanzania and Kenya, a journey that can be 1,900 miles (3,000 km) long. In the dry season, the wildebeest move from acacia woodlands onto the Serengeti Plain in search of water and grass. They return in the wet season.

Calves are born at the beginning of the rainy season, and stay with their mothers until eight months of age, when they leave to form peer-group herds.

Wildebeest never stay in one spot long enough to damage the environment, and as the columns of animals move forward, the soil is fertilized by their droppings.

Zebras migrate across the Serengeti at the same time, eating grasses cropped short by the wildebeest, and using the tens of thousands of animals as protection from predators.

River crossings, such as the Mara River, can be deadly, with crocodiles and other predators lying in wait for an animal to stumble or drift away from the mass.

True Cattle, Goats and Sheep

Location

■ True cattle
■ Goats and sheep
■ True cattle, goats and sheep

YAK
Bos grunniens

MOUNTAIN ANOA
Bubalus quarlesi

AMERICAN BISON
Bison bison

EUROPEAN BISON
Bison bonasus

AFRICAN BUFFALO
Syncerus caffer

DOMESTIC BREEDS OF CATTLE

Hereford cattle

Friesian cattle

Galloway cattle

Brahmin cattle

Ankole cattle

Texas long-horned cattle

MUSKOX
Ovibos moschatus

MOUFLON
Ovis orientalis musimon

DALL'S SHEEP
Ovis dalli

BIGHORN SHEEP
Ovis canadensis

HIMALAYAN TAHR
Hemitragus jemlahicus

WILD GOAT
Capra aegagrus

MOUNTAIN GOAT
Oreamnos americanus

DOMESTIC BREEDS OF SHEEP AND GOATS

Leicester sheep

Cashmere goat

Lincoln sheep

Black-faced sheep

Merino sheep

Domestic sheep

Romney sheep

Rhinoceroses

Location

■ Rhinoceroses

RHINO DEFENSE

Groups of rhinoceroses form a circle facing outward to protect their calves.

BLACK RHINO CHARGING

Black rhinos attack animals that threaten them.

Horns made of hollow filaments similar to hair

Spotted hyena

Black rhinos use their horn to gore and toss would-be attackers.

JAVAN RHINOCEROS
Rhinoceros sondaicus

INDIAN RHINOCEROS
Rhinoceros unicornis

SUMATRAN RHINOCEROS
Dicerorhinus sumatrensis

WHITE RHINOCEROS
Ceratotherium simum

BLACK RHINOCEROS
Diceros bicornis

RHINO COMPARISON

Indian rhinoceros
Asian one-horned

Black rhinoceros
African two-horned

Horses, Asses and Zebras

HORSE GALLOPING
When horses gallop, each hoof strikes the ground separately.

HORSE SKELETON AND DIGESTION
Horses chew their food once only, and have a single stomach with a long intestinal system and large, long teeth to break down tough, fibrous grasses.

Rectum

Pelvic bone

Lower vertebrae

Dorsal colon

Ventral colon

Scapula

Esophagus

Stomach

Humerus

Elbow

Tibia

Anklebone

Carpal

Cecum

Small intestine

Hoof (toenail)

HORSE SKULL

Postorbital bar

Eye socket behind teeth

Prominent nasal bones

Incisors for snapping stalks

Strong jawbone for chewing plant matter

Large molars for grinding cellulose

FLEHMEN RESPONSE
Horses often curl up their top lip to heighten their sense of smell, especially when males determine a female's fertility.

Prehensile upper lip

ZEBRA DEFENSE
Zebras will kick with their sharp hooves and strong rear thigh muscles to defend themselves against predators. Blows to the head can kill a would-be attacker.

Horses, Asses and Zebras

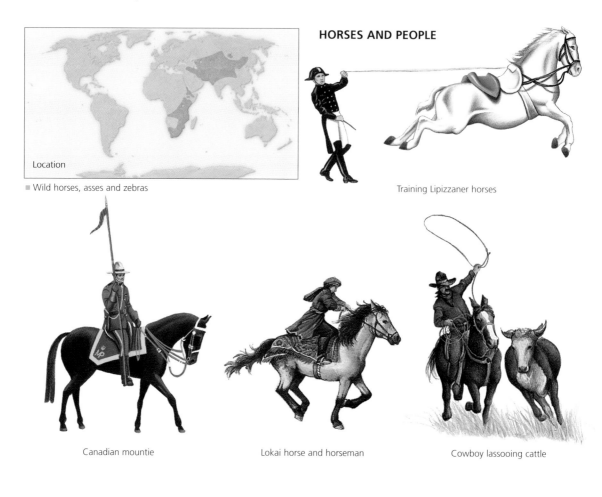

Location

■ Wild horses, asses and zebras

HORSES AND PEOPLE

Training Lipizzaner horses

Canadian mountie

Lokai horse and horseman

Cowboy lassooing cattle

Playing polo

Racing horses

Austrian sled racing

204

KIANG
Equus kiang

ONAGER
Equus onager

ASS
Equus asinus

HORSE
Equus caballus

TARPAN
Equus caballus gmelini

ZEBRA
Equus sp.

DOMESTIC HORSES

Feral horses

Plains pony

Arabian horse

Aardvark, Hyraxes and Tapirs

Location

■ Aardvark and hyraxes
■ Tapirs

YELLOW-SPOTTED HYRAX
Heterohyrax brucei

ROCK HYRAXES
Procavia capensis

AARDVARK
Orycteropus afer

AARDVARK ANATOMY

Tops of ears can fold back to protect them when digging.

Aardvarks sniff out ants and termites to eat.

MOUNTAIN TAPIR
Tapirus pinchaque

BAIRD'S TAPIR
Tapirus bairdii

**SOUTH AMERICAN TAPIR
WITH YOUNG**
Tapirus terrestris

MALAYAN TAPIR
Tapirus indicus

Elephants

ELEPHANT SKIN

Cross-section of hair

Epidermis

Nerve endings

Hair follicle

Dermis

Subcutaneous layer

Muscle layer

ELEPHANT FOOT
Elephants walk on toes of feet.

Fibrous pads support the toes.

ASIAN ELEPHANT

Domed forehead

Ears shorter than neck

Convex or level back

One finger on tip of trunk

Skin relatively smooth

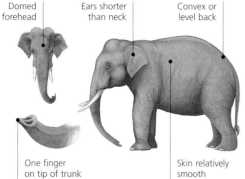

AFRICAN ELEPHANT

No dome on forehead

Ears longer than neck

Concave back

Two fingers on tip of trunk

Skin wrinkled and baggy

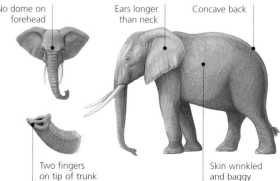

ELEPHANT GROWTH

Less than one year

One year

Three years

Six years

10 years

15 years

40 years

AFRICAN ELEPHANT

AFRICAN ELEPHANT SKELETON

Neck vertebrae

Scapula

Ribcage

Lower vertebrae

Skull

Eye socket

Teeth

Tusk

Humerus

Trunk

Pelvic bone

Tail

Femur

Forefoot

Hindfoot

Elephants

Location

■ Elephants

ELEPHANT TUSKS

Tusks have cavities
containing nerve
endings, which are
pressure-sensitive.

AFRICAN BUSH ELEPHANTS
Loxodonta africana

SUMATRAN ASIAN ELEPHANTS
Elephas maximus sumatranus

SRI LANKAN ASIAN ELEPHANTS
Elephas maximus maximus

MAINLAND ASIAN ELEPHANTS
Elephas maximus indicus

FOREST AFRICAN ELEPHANTS
Loxodonta cyclotis

Dugong and Manatees

Location

- Dugong
- Manatees

MANATEE FEEDING

Manatees can close their nostrils under water.

Skin is thick and tough.

Flexible lips guide plants into mouth.

Flippers enable manatee to hold water plants while feeding.

AFRICAN MANATEE
Trichechus senegalensis

DUGONG
Dugong dugon

WEST INDIAN MANATEE
Trichechus manatus

AMAZONIAN MANATEE
Trichechus inunguis

Elephant Shrews

Location

■ Elephant shrews

ELEPHANT SHREW HEAD

Ear

Trunklike nose

Large, white-rimmed eyes

ROCK ELEPHANT SHREW HABITAT
Elephantulus myurus

Rock elephant shrews nest in the crevices of rock overhangs to protect themselves from predators and the elements.

They have large eyes and a long, sensitive and mobile snout, which is used to find insects.

Rock elephant shrews are generally soft gray-brown in color, which acts as camouflage from predators.

SHORT-EARED ELEPHANT SHREW
Macroscelides proboscideus

GOLDEN-RUMPED ELEPHANT SHREW
Rhynchocyon chrysopygus

RUFOUS ELEPHANT SHREW
Elephantulus rufescens

NORTH AFRICAN ELEPHANT SHREW
Elephantulus rozeti

FOUR-TOED ELEPHANT SHREW
Petrodromus tetradactylus

Rodents

EVOLUTION OF RODENTS

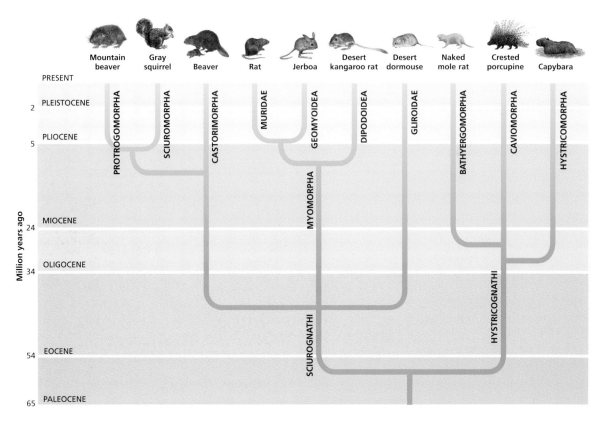

Mountain beaver	Gray squirrel	Beaver	Rat	Jerboa	Desert kangaroo rat	Desert dormouse	Naked mole rat	Crested porcupine	Capybara	

PRESENT

PLEISTOCENE

2

PLIOCENE

5

MIOCENE

24

OLIGOCENE

34

EOCENE

54

PALEOCENE

65

Million years ago

PROTROGOMORPHA

SCIUROMORPHA

CASTORIMORPHA

MURIDAE

GEOMYOIDEA

DIPODOIDEA

GLIROIDAE

BATHYERGOMORPHA

CAVIOMORPHA

HYSTRICOMORPHA

MYOMORPHA

SCIUROGNATHI

HYSTRICOGNATHI

RODENT SKULL

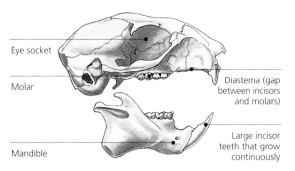

Eye socket

Molar

Mandible

Diastema (gap between incisors and molars)

Large incisor teeth that grow continuously

RODENT HEAD

Ear

Cheeks able to store food while gathering

Mandible

Excellent vision

Whiskers

Lower incisor

RODENT JAW MUSCLES

■ Lateral jaw muscles ■ Deep jaw muscles

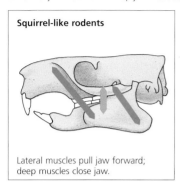

Squirrel-like rodents

Lateral muscles pull jaw forward; deep muscles close jaw.

Mouselike rodents

Lateral and deep muscles work together to pull jaw forward.

Cavylike rodents

Lateral muscles close jaw; deep muscles pull jaw forward.

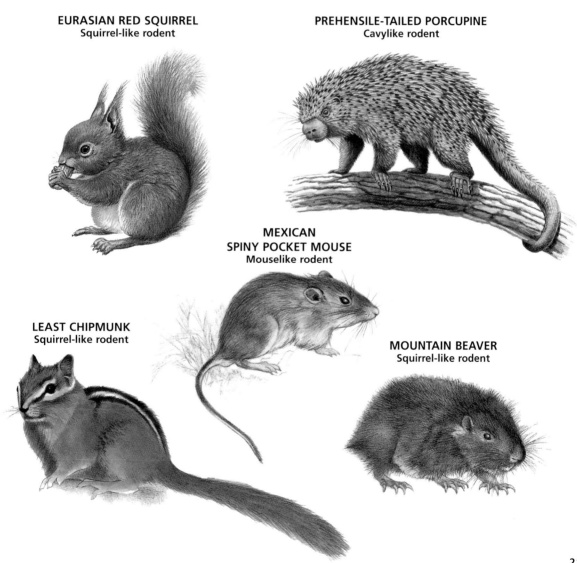

EURASIAN RED SQUIRREL
Squirrel-like rodent

PREHENSILE-TAILED PORCUPINE
Cavylike rodent

MEXICAN
SPINY POCKET MOUSE
Mouselike rodent

LEAST CHIPMUNK
Squirrel-like rodent

MOUNTAIN BEAVER
Squirrel-like rodent

Squirrel-like Rodents

Location

■ Squirrel-like rodents

PRAIRIE DOGS
Cynomys sp.

PRAIRIE DOG COMMUNITY

Prairie dogs greet each other with kissing ceremonies.

Guards whistle an alarm when in danger.

Ferrets can crawl through the tunnels in search of prey.

Females build grass nests safe underground.

Members of different coteries greet each other by sniffing.

Prairie dogs live in social units called coteries. A number of interconnected coteries form a town.

Snakes and burrowing owls often take over unused burrows.

EASTERN FOX SQUIRREL
Sciurus niger

DOUGLAS' SQUIRREL
Tamiasciurus douglasii

MARMOT
Marmota sp.

WOODCHUCK
Marmota monax

GRAY SQUIRREL
Sciurus sp.

ABERT'S SQUIRREL
Sciurus aberti

EASTERN CHIPMUNK
Tamias striatus

Squirrel-like Rodents

BLACK GIANT SQUIRREL
Ratufa bicolor

DELMARVA FOX SQUIRREL
Sciurus niger cenerus

AMERICAN RED SQUIRREL
Tamiasciurus hudsonicus

LEAST CHIPMUNK
Tamias minimus

HOARY MARMOT
Marmota caligata

COLUMBIAN GROUND SQUIRREL
Spermophilus columbianus

FLYING SQUIRREL ANATOMY

Spurs on the forefeet and hindfeet support the membrane.

Membrane extends like a parachute, the squirrel softly gliding to the ground.

Abdomen

Hindfoot

Gliding membrane extends from foreleg to hindleg.

Rudderlike tail lets squirrel maneuver through the air.

EURASIAN RED SQUIRREL
Sciurus vulgaris

SOUTHERN FLYING SQUIRREL
Glaucomys volans

Beavers

Location

■ Beavers and mountain beavers

AMERICAN BEAVER AND KIT
Castor canadensis

BEAVER DAM AND LODGE

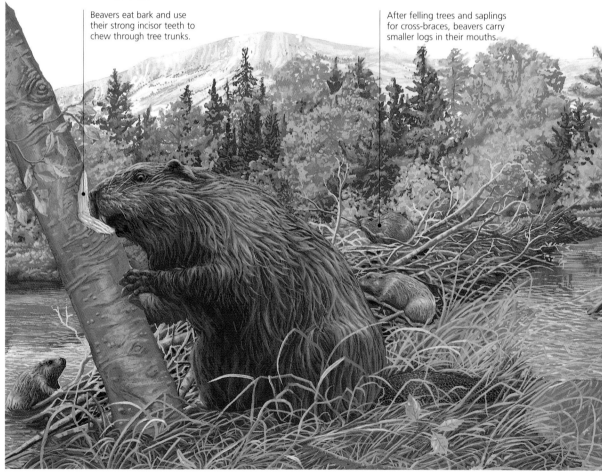

Beavers eat bark and use their strong incisor teeth to chew through tree trunks.

After felling trees and saplings for cross-braces, beavers carry smaller logs in their mouths.

EURASIAN BEAVER
Castor fiber

Beaver lodge

The lodge's entrance is underwater to protect it from predators.

Inside the lodge, temperatures are much more constant than in the external climate.

DAM CONSTRUCTION
Heavier logs run crosswise.

Smaller logs run lengthwise.

LODGE ENTRANCE
Before damming

After damming

Mouselike Rodents

Location

■ Mouselike rodents

MUSKRAT
Ondatra zibethicus

MASKED MOUSE-TAILED DORMOUSE
Myomimus personatus

DESERT DORMOUSE
Selevinia betpakdalaensis

**WOODLAND
DORMOUSE**
Graphiurus murinus

CRESTED RAT
Lophiomys imhausi

LEMMING
Lemmus sp.

BLACK RAT
Rattus mindorensis

HOUSE RAT
Rattus rattus

**LONG-TAILED
POCKET MOUSE**
Chaetodipus formosus

GERBIL
Gerbillus sp.

JERBOA
Jaculus sp.

BANK VOLE
Clethrionomys glareolus

GRASSHOPPER MOUSE
Onychomys sp.

DESERT KANGAROO RAT
Dipodomys deserti

Cavylike Rodents

Location

■ Cavylike rodents

BRAZILIAN PORCUPINE
Coendou prehensilis

CAPYBARA
Hydrochaeris hydrochaeris

BRAZILIAN AGOUTI
Dasyprocta leporina

CRESTED PORCUPINE
Hystrix cristata

MALAYAN PORCUPINE
Hystrix brachyura

CORURO
Spalacopus cyanus

GUINEA PIG
Cavia aperea

CHILEAN ROCK RAT
Aconaemys fuscus

PLAINS VISCACHA
Lagostomus maximus

CHINCHILLA
Chinchilla lanigera

PATAGONIAN MARA
Dolichotis patagonum

SPINY RAT
Family *Echimyidae*

Rabbits, Hares and Pikas

BLACK-LIPPED PIKA
Ochotona curzoniae
Pikas are short-legged and short-eared relatives of rabbits and hares. Mainly herbivorous, they live on the steppes and plains of North America and continental Asia.

RABBIT ANATOMY

Long ears to detect approach of predators and to keep cool

Curved skull

Eyes at side of head to locate predators from above and behind

Continuously growing incisors

RABBIT WARREN

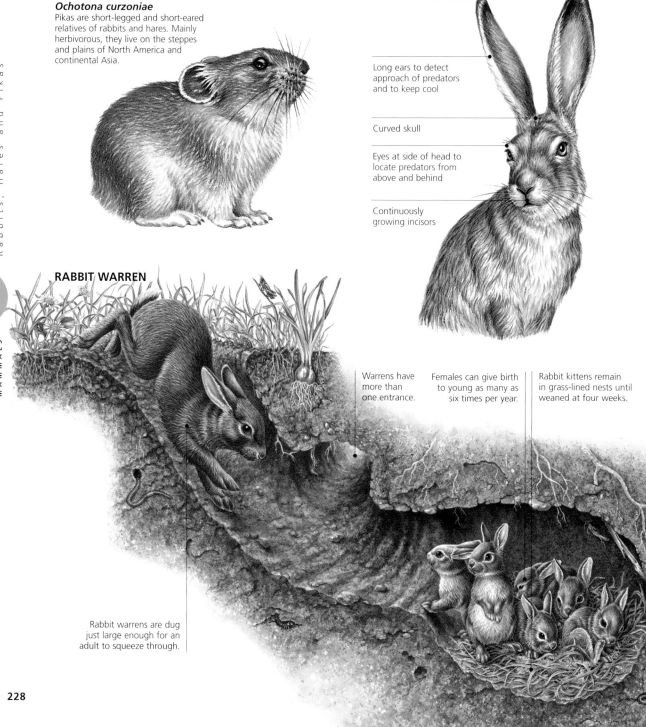

Warrens have more than one entrance.

Females can give birth to young as many as six times per year.

Rabbit kittens remain in grass-lined nests until weaned at four weeks.

Rabbit warrens are dug just large enough for an adult to squeeze through.

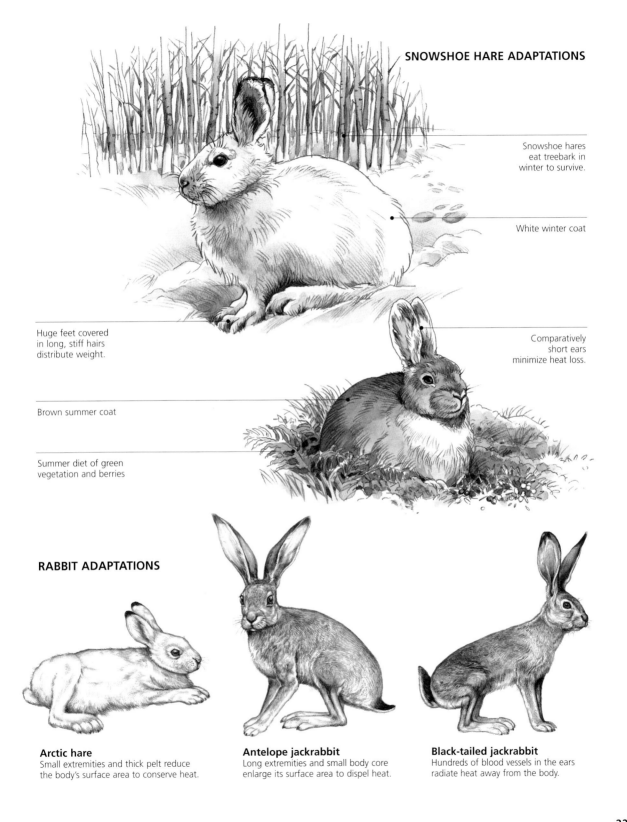

SNOWSHOE HARE ADAPTATIONS

Snowshoe hares eat treebark in winter to survive.

White winter coat

Huge feet covered in long, stiff hairs distribute weight.

Comparatively short ears minimize heat loss.

Brown summer coat

Summer diet of green vegetation and berries

RABBIT ADAPTATIONS

Arctic hare
Small extremities and thick pelt reduce the body's surface area to conserve heat.

Antelope jackrabbit
Long extremities and small body core enlarge its surface area to dispel heat.

Black-tailed jackrabbit
Hundreds of blood vessels in the ears radiate heat away from the body.

Rabbits, Hares and Pikas

Location

■ Rabbits and hares
■ Rabbits, hares and pikas

SUMATRAN RABBIT
Nesolagus netscheri

RYUKYU RABBIT
Pentalagus furnessi

VOLCANO RABBIT
Romerolagus diazi

BRUSH RABBIT
Sylvilagus bachmani

RIVENNE RABBIT
Bunolagus monticularis

PYGMY RABBIT
Brachylagus idahoensis

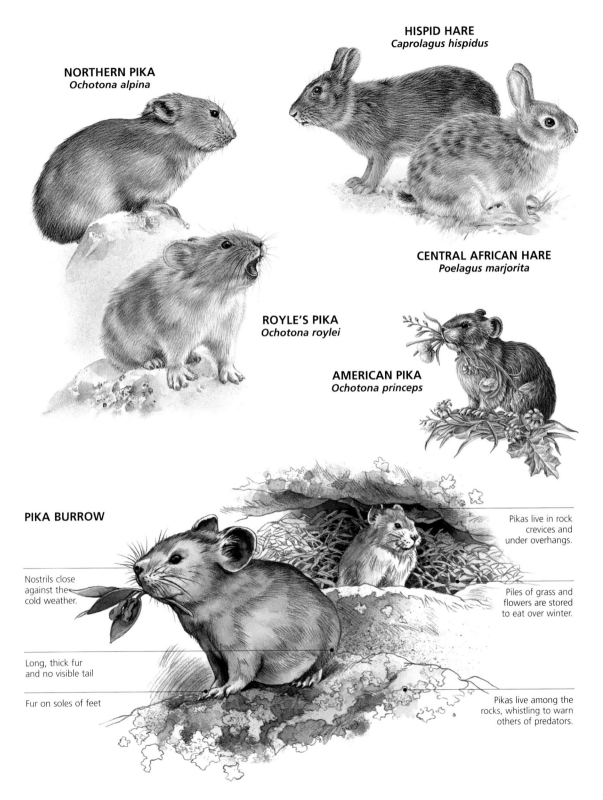

NORTHERN PIKA
Ochotona alpina

HISPID HARE
Caprolagus hispidus

ROYLE'S PIKA
Ochotona roylei

CENTRAL AFRICAN HARE
Poelagus marjorita

AMERICAN PIKA
Ochotona princeps

PIKA BURROW

Nostrils close against the cold weather.

Long, thick fur and no visible tail

Fur on soles of feet

Pikas live in rock crevices and under overhangs.

Piles of grass and flowers are stored to eat over winter.

Pikas live among the rocks, whistling to warn others of predators.

231

Classifying Birds

EVOLUTION OF BIRDS

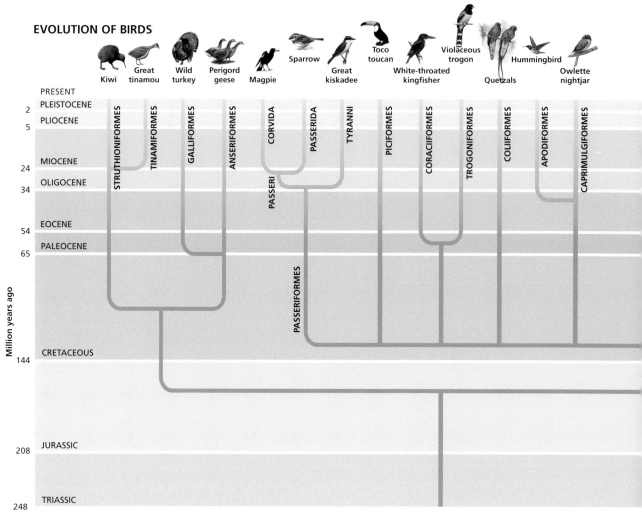

Kiwi · Great tinamou · Wild turkey · Perigord geese · Magpie · Sparrow · Great kiskadee · Toco toucan · White-throated kingfisher · Violaceous trogon · Quetzals · Hummingbird · Owlette nightjar

Million years ago

PRESENT
PLEISTOCENE
2
PLIOCENE
5
MIOCENE
24
OLIGOCENE
34
EOCENE
54
PALEOCENE
65
CRETACEOUS
144
JURASSIC
208
TRIASSIC
248

STRUTHIONIFORMES · TINAMIFORMES · GALLIFORMES · ANSERIFORMES · CORVIDA · PASSERI · PASSERIDA · TYRANNI · PASSERIFORMES · PICIFORMES · CORACIIFORMES · TROGONIFORMES · COLIIFORMES · APODIFORMES · CAPRIMULGIFORMES

Male (top) and female gambel's quails

Bonellis eagle

Painted stork

236

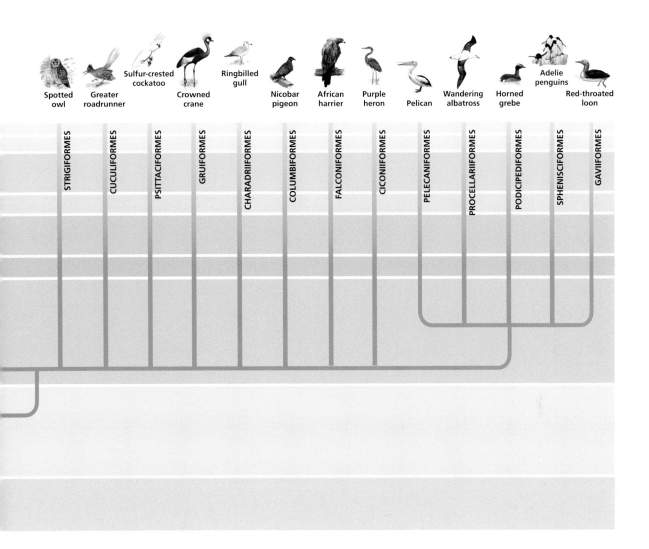

Spotted owl

Greater roadrunner

Sulfur-crested cockatoo

Crowned crane

Ringbilled gull

Nicobar pigeon

African harrier

Purple heron

Pelican

Wandering albatross

Horned grebe

Adelie penguins

Red-throated loon

STRIGIFORMES

CUCULIFORMES

PSITTACIFORMES

GRUIFORMES

CHARADRIIFORMES

COLUMBIFORMES

FALCONIFORMES

CICONIIFORMES

PELECANIFORMES

PROCELLARIIFORMES

PODICIPEDIFORMES

SPHENISCIFORMES

GAVIIFORMES

Oilbird

Tree swallow

Cedar waxwing

Bird Characteristics

TYPES OF FEATHERS

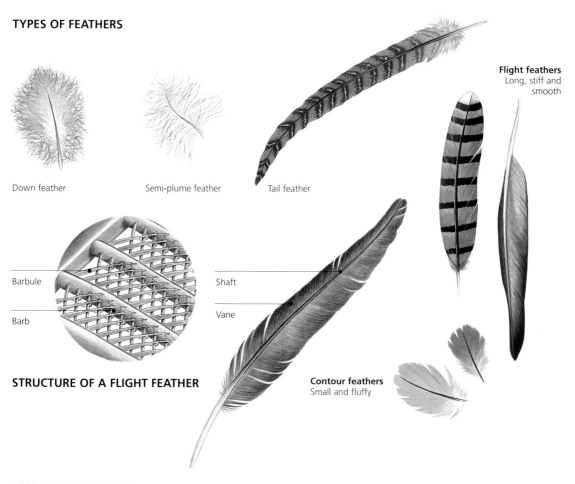

Down feather

Semi-plume feather

Tail feather

Flight feathers
Long, stiff and smooth

Barbule

Barb

Shaft

Vane

STRUCTURE OF A FLIGHT FEATHER

Contour feathers
Small and fluffy

REPLACING FEATHERS

Worn feather

New feather pushes out the old.

Feather unfurls from shaft.

New feather

PLUMAGE FEATHERS

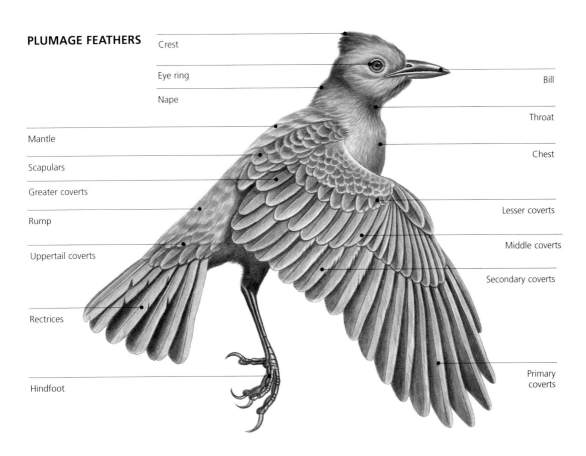

Crest

Eye ring

Nape

Mantle

Scapulars

Greater coverts

Rump

Uppertail coverts

Rectrices

Hindfoot

Bill

Throat

Chest

Lesser coverts

Middle coverts

Secondary coverts

Primary
coverts

INSIDE AN EGG

Newly laid egg
Small embryo
uses yolk and
albumen as food.

Embryonic chick
As embryo grows,
wastes are stored
in a special sac.

Developing chick
When close to
hatching, chick
nearly fills egg.

Ready to hatch
Unhatched chicks
use an egg tooth
to break free.

Evolution of Birds

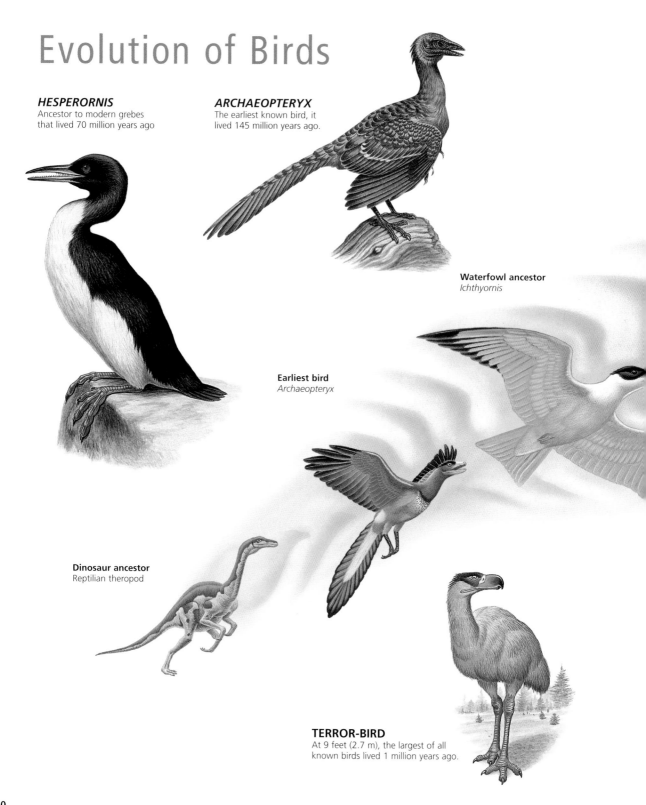

HESPERORNIS
Ancestor to modern grebes
that lived 70 million years ago

ARCHAEOPTERYX
The earliest known bird, it
lived 145 million years ago.

Waterfowl ancestor
Ichthyornis

Earliest bird
Archaeopteryx

Dinosaur ancestor
Reptilian theropod

TERROR-BIRD
At 9 feet (2.7 m), the largest of all
known birds lived 1 million years ago.

EVOLUTION OF MODERN BIRDS

Modern waterfowl
Snow goose

DINORNIS MAXIMUS
Recently extinct flightless bird
that lived 200–300 years ago

TERATORNIS MERRIAMI
An ancestor of modern vultures,
it lived 2 million years ago.

241

Bird Anatomy

AVIAN ANATOMY

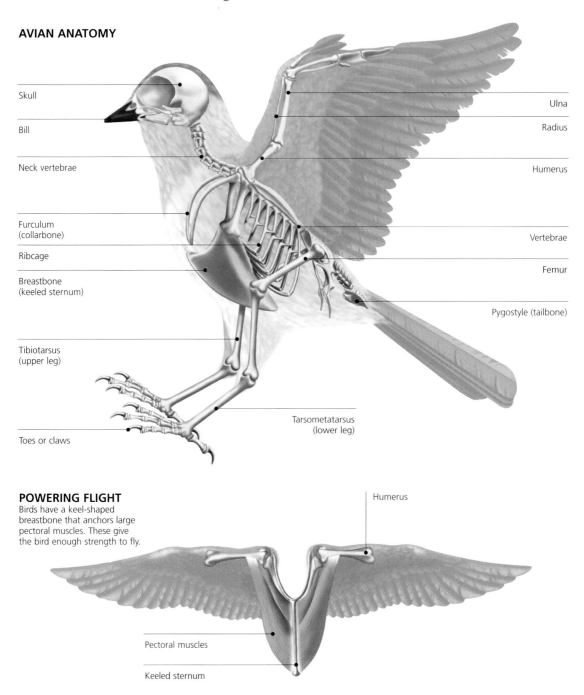

Skull

Bill

Neck vertebrae

Furculum
(collarbone)

Ribcage

Breastbone
(keeled sternum)

Tibiotarsus
(upper leg)

Toes or claws

Ulna

Radius

Humerus

Vertebrae

Femur

Pygostyle (tailbone)

Tarsometatarsus
(lower leg)

POWERING FLIGHT

Birds have a keel-shaped
breastbone that anchors large
pectoral muscles. These give
the bird enough strength to fly.

Humerus

Pectoral muscles

Keeled sternum

DIGESTIVE SYSTEM

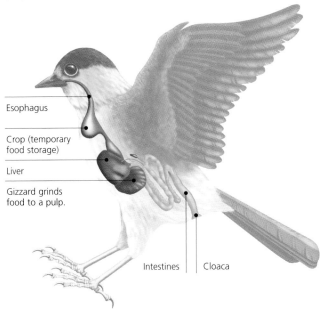

Esophagus

Crop (temporary food storage)

Liver

Gizzard grinds food to a pulp.

Intestines

Cloaca

HOLLOW BONES

Most bones in a bird's body are thin-walled and hollow, with struts and braces providing maximum strength for minimum weight.

Air space

Strut

LUNG CROSS-SECTION

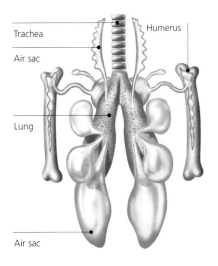

Trachea

Air sac

Humerus

Lung

Air sac

LEG MUSCLES

All birds have powerful leg muscles in the top of the leg, near their center of gravity. These are connected to the toes by long tendons.

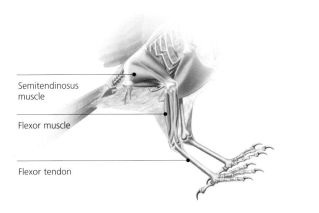

Semitendinosus muscle

Flexor muscle

Flexor tendon

HEART CROSS-SECTION

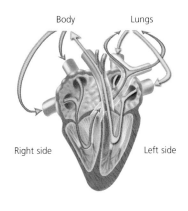

Body

Lungs

Right side

Left side

Flightless Birds

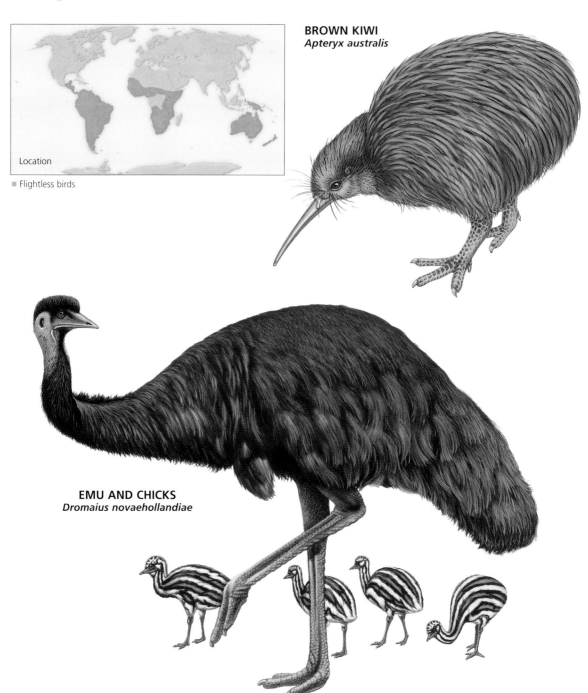

Location

■ Flightless birds

BROWN KIWI
Apteryx australis

EMU AND CHICKS
Dromaius novaehollandiae

Female great tinamou head

Emu at rest

Emu egg

Great tinamou

Greater rhea

Rhea foot

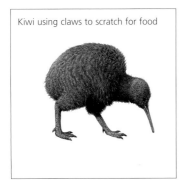
Kiwi using claws to scratch for food

Ostrich tail feather

Ostrich

Kiwi head and bill

Cassowary foot

Southern cassowary

Penguins

Location

■ Penguins

EMPEROR PENGUIN ANATOMY

Feathers

Foot

| Scaly, oily tips | Bend at the base | Fluffy down | Ankle feathers | Long toenails to grip the ice |

EMPEROR PENGUINS SWIMMING

Emperor penguins are swift, agile swimmers, and can swim up to 20 miles (32 km) per hour. They dive through a hole in the ice and use their stiff, flipperlike wings to chase their prey.

246

FJORDLAND PENGUIN
Eudyptes pachyrhynchus

ROCKHOPPER PENGUIN
Eudyptes chrysocome

MAGELLANIC PENGUIN
Spheniscus magellanicus

FAIRY PENGUIN
Eudyptula minor

EMPEROR PENGUIN
Aptenodytes forsteri

CHINSTRAP PENGUIN
Pygoscelis antarctica

YELLOW-EYED PENGUIN
Megadyptes antipodes

Grebes and Divers

Location

- Grebes and dabchicks
- Divers (loons)
- Grebes, dabchicks and divers

GREAT CRESTED GREBE
Podiceps cristatus

RED-NECKED GREBE
Podiceps grisegena

PIED-BILLED GREBE WITH CHICKS
Podilymbus podiceps

Great northern

Red-throated

Black-throated

GROUP OF DIVERS (LOONS)
Gavia sp.

COMMON LOON
Gavia immer

EARED GREBE
Podiceps nigricollis

HORNED GREBE
Podiceps auritus

248

WESTERN GREBES 'RUSHING'

Rushing involves two or more grebes
running upright across the water as
part of a ritualistic courtship display.

Lakeside vegetation

Wings are drawn
back and flexed.

Necks are arched with
head slightly bowed.

Males use this display to
defend their territory.

RED-THROATED LOON
Gavia stellata

LITTLE GREBE
Tachybaptus ruficollis

Pacific (left) and common loon heads

Western (top) and Clark's grebe heads

Yellow-billed loon head and neck

Albatrosses and Petrels

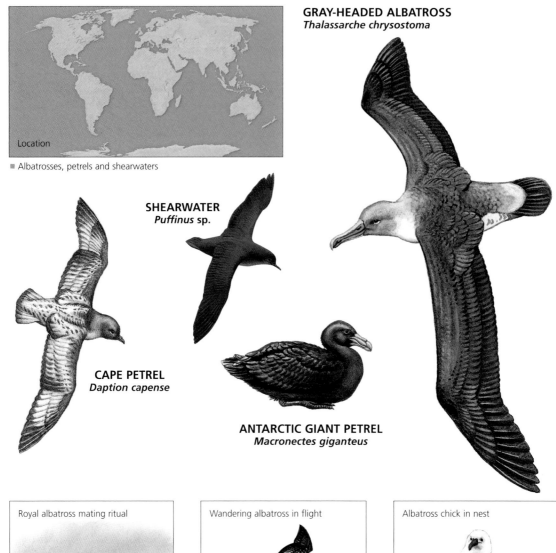

Location

■ Albatrosses, petrels and shearwaters

SHEARWATER
Puffinus sp.

GRAY-HEADED ALBATROSS
Thalassarche chrysostoma

CAPE PETREL
Daption capense

ANTARCTIC GIANT PETREL
Macronectes giganteus

Royal albatross mating ritual

Wandering albatross in flight

Albatross chick in nest

SHORT-TAILED SHEARWATER MIGRATION

June–August
Shearwaters spend the non-breeding season in the North Pacific Ocean.

April–May
Chick reaches adult weight and grows flight feathers, ready to leave the nest.

Location

■ Breeding ■ Wintering

September
It takes several weeks for the birds migrating south to reach their nesting colonies, along the coast of Australia.

February–March
The young chick is fed every three days as its parents alternate between care and hunting.

Migratory path

October–November
After courting, shearwaters establish pair bonds for the year, and repair their nest site.

December–January
Both parents incubate the egg, laid late November, until it hatches in January.

251

Pelicans

Location

■ Pelicans

BROWN PELICAN
Pelecanus occidentalis

Shag with summer plumage

British cormorant spreading wings

Blue-footed booby on rock

Magnificent frigatebird with throat sac

Gannet at rest

Australian pelican

DOUBLE-CRESTED CORMORANT
Phalacrocorax auritus

AMERICAN WHITE PELICAN
Pelecanus erythrorhynchos

BRANDT'S CORMORANT
Phalacrocorax penicillatus

BROWN PELICAN HEADS

Summer plumage

Winter plumage

PELICAN DIVING FOR FISH

Eyes remain open underwater.

Pouch scoops up fish and water.

Fish caught are carried in the esophagus, not the pouch, before swallowing.

All pelicans have webbing that runs between the toes.

253

Herons

Location

■ Herons

Juvenile   Male   Female

EURASIAN SPOONBILLS
Platalea leucorodia

Adult

Juvenile

BLACK-CROWNED NIGHT HERONS
Nycticorax nycticorax

GREAT BITTERN
Botaurus stellaris

**SHOEBILL
WITH YOUNG**
Balaeniceps rex

LITTLE EGRET
Egretta garzetta

**GREAT WHITE
EGRET**
Egretta alba

Yellow-crowned night heron head

Roseate spoonbill spreading wings

Roseate spoonbill at rest

Juvenile little blue heron

Great blue heron in flight

African spoonbill head

Black stork wading

Great blue heron head

Scarlet ibis and skeleton

White stork in flight

Great bittern camouflaged in reeds

Great (left) and little bitterns in flight

Herons

GREEN HERONS FISHING
The green herons of Florida, U.S.A., have learnt that dropping breadcrumbs into the water attracts fish. The next generation are then taught the same tactic.

GRAY HERON
Ardea cinerea

STRIATED HERON
Butorides striatus

GREEN HERON
Butorides virescens

WALDRAPP
Geronticus eremita

LITTLE BLUE HERON
Egretta caerulea

CATTLE EGRET
Bubulcus ibis

PURPLE HERON
Ardea purpurea

GOLIATH HERON
Ardea goliath

PAINTED STORK
Mycteria leucocephala

ROSEATE SPOONBILL
Ajaia ajaja

257

Flamingos

Location

■ Flamingos

LESSER FLAMINGO FEEDING

Long, flexible neck bends to the water while the flamingo remains upright.

Upper jaw contains a row of slits, called lamellae, that act like a sieve.

Tongue and lower jaw strain food from the water through the lamellae.

LESSER FLAMINGO
Phoenicopterus minor

PUNA FLAMINGO
Phoenicopterus jamesi

Flamingo at rest

Flamingo feeding

Greater flamingos

GREATER FLAMINGO COLONY

Flamingos fly with their neck stretched out and long legs trailing behind.

Flamingos are highly social birds that live and breed in large lakeside colonies.

Their dramatic pink coloring is thought to come from certain algae that they eat.

A hooked bill scoops up water containing tiny animals and algae.

Waterfowl

Location

■ Waterfowl

COMMON MOORHENS
Gallinula chloropus

BARNACLE GEESE MIGRATING
Branta leucopsis
Barnacle geese fly in a characteristic
v-shaped formation when migrating
to and from Greenland and Europe.

Lead bird flies at the head of
the v-shape, breaking up air
currents for those that follow.

Migrating birds spend a short
time as lead bird, before
dropping back into the flock.

Canada goose with goslings

Velvet scoters

Gaggle of Perigord geese

Mallard foot showing webbed toes

Mallard in flight

Mallard diving for fish

Ruddy duck diving

Dark- and pale-bellied Brant geese

Black swan swimming

Northen shoveler with spatulate bill

Harlequin ducks swimming

Whooper and Bewick's swans

Waterfowl

RED-BREASTED GOOSE
Branta ruficollis

KING EIDER
Somateria spectabilis

WOOD DUCK
Aix sponsa

NORTHERN SCREAMER
Chauna chavaria

GRAYLAG GEESE
Anser anser

COMMON GOLDENEYES
Bucephala clangula

Male

Female

SNOW GOOSE
Chen caerulescens

AMERICAN BLACK DUCK
Anas rubripes

CINNAMON TEAL
Anas cyanoptera

Adult

Juvenile

MUTE SWANS
Cygnus olor

FEMALE COMMON EIDERS WITH DUCKLINGS
Somateria mollissima

Birds of Prey

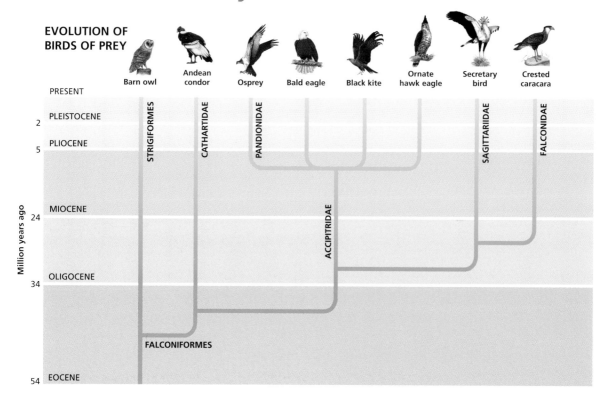

EVOLUTION OF BIRDS OF PREY

Barn owl   Andean condor   Osprey   Bald eagle   Black kite   Ornate hawk eagle   Secretary bird   Crested caracara

Million years ago

PRESENT

PLEISTOCENE

2

PLIOCENE

5

MIOCENE

24

OLIGOCENE

34

EOCENE

54

STRIGIFORMES   CATHARTIDAE   PANDIONIDAE   ACCIPITRIDAE   SAGITTARIIDAE   FALCONIDAE

FALCONIFORMES

WING SPAN COMPARISON

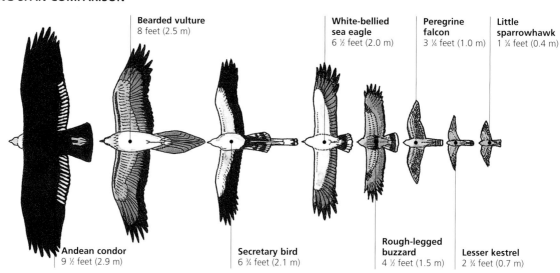

Bearded vulture
8 feet (2.5 m)

White-bellied sea eagle
6 ½ feet (2.0 m)

Peregrine falcon
3 ¼ feet (1.0 m)

Little sparrowhawk
1 ¼ feet (0.4 m)

Andean condor
9 ½ feet (2.9 m)

Secretary bird
6 ¾ feet (2.1 m)

Rough-legged buzzard
4 ½ feet (1.5 m)

Lesser kestrel
2 ¾ feet (0.7 m)

RAPTOR ANATOMY

Primary wing feathers

Alula

Upperwing coverts

Forward-facing eyes

Hooked beak overhangs mandible.

Underwing coverts

Auxiliary feathers

Strong toes to lift and carry prey

Sharp, curved talons to tear flesh

Alula helps prevent stalling when banking sharply.

Primary wing feathers

Secondary wing feathers

Undertail coverts

Raptors have 12 or 14 tail feathers.

RAPTOR HEADS AND CLAWS

Osprey

White-backed vulture

Eurasian sparrowhawk

Harpy eagle

Osprey: fish eater

Vuture: carrion eater

Sparrowhawk: bird eater

Harpy eagle: mammal eater

Eagles, Kites and Hawks

Location

- Eagles, kites and hawks
- Eagles, kites, hawks, harriers, serpent eagles and osprey

SWALLOW-TAILED KITE
Elanoides forficatus

WHITE-BELLIED SEA-EAGLE
Haliaeetus leucogaster

BRAHMINY KITE
Haliastur indus

BIRDS

HOOK-BILLED KITE
Chondrohierax uncinatus

SNAIL KITE
Rostrhamus sociabilis

CRESTED SERPENT-EAGLE
Spilornis cheela

BLACK-SHOULDERED KITE
Elanus caeruleus

GREAT PHILIPPINE EAGLE
Pithecophaga jefferyi

267

Eagles, Kites and Hawks

PEARL KITE
Gampsonyx swainsonii

SLATE-COLORED HAWK
Leucopternis schistacea

ROUGH-LEGGED HAWK
Buteo lagopus

OSPREY
Pandion haliaetus

ORNATE HAWK EAGLE
Spizaetus ornatus

AFRICAN HARRIER-HAWK
Polyboroides typus

PACIFIC BAZA
Aviceda subcristata

**EASTERN CHANTING
GOSHAWK**
Melierax poliopterus

BATELEUR
Terathopius ecaudatus

269

Falcons

Location

- Falcons and relatives
- Falcons, relatives and caracaras

AMERICAN KESTREL
Falco sparverius

GREATER KESTREL
Falco rupicoloides

STRIATED CARACARA
Phalcoboenus australis

SPOT-WINGED FALCONET
Spiziapteryx circumcinctus

Lined forest falcon

Peregrine falcon hunting pigeon

American kestrel displaying plumage

PEREGRINE FALCON
Falco peregrinus

EURASIAN HOBBY
Falco subbuteo

PYGMY FALCON
*Polihierax
semitorquatus*

**PEREGRINE FALCONS
MATING**
The male falcon alights on the
female's lowered back, and balances
by flapping furiously. He lowers his
tail under hers so that their cloacas
meet and sperm is exchanged.

CRESTED CARACARA
Caracara plancus

271

Secretary Bird and Vultures

Location

- Old World vultures
- New World vultures
- Old World vultures and secretary bird

ANDEAN CONDOR
Vultur gryphus

PALM-NUT VULTURE
Gypohierax angolensis

SECRETARY BIRD
Sagittarius serpentarius

LAMMERGEIER
Gypaetus barbatus

Lappet-faced vulture spreading wings

Griffon vulture on rock

Secretary bird grabbing snake with feet

Lappet-faced vulture

Juvenile lammergeier

Black vulture head

Eurasian griffon

Juvenile (left) and adult turkey vultures

Turkey vulture in flight

Gamebirds

Location

- Gamebirds

MALEO
Macrocephalon maleo

MALLEEFOWL NEST

Sand layer makes the nest warmer or cooler.

Leaf litter layer slowly rots to generate heat.

Eggs incubate at the constant temperature of about 91° F (33° C).

Malleefowls eat herbs in winter, fruits and seeds in summer.

They do not drink water, obtaining all necessary liquids from their solid diet.

Male tests temperature inside the nest with his mouth and tongue.

Chicken egg

Wild turkey displaying plumage

Baby fowl and egg

Gray partridges preparing to dust bath

Ptarmigan with winter plumage

Ptarmigan with summer plumage

Ruffed grouse displaying tail feathers

Ring-necked pheasant skeleton

Ring-necked pheasant

Temminck's tragopan

Satyr tragopan

Capercaillie displaying tail feathers

Gamebirds

INDIAN PEAFOWL
Pavo cristatus

HOATZIN
Opisthocomus hoazin

WHITE-CRESTED GUAN
Penelope pileata

GREATER PRAIRIE-CHICKEN
Tympanuchus cupido

CALIFORNIA QUAIL
Callipepla californica

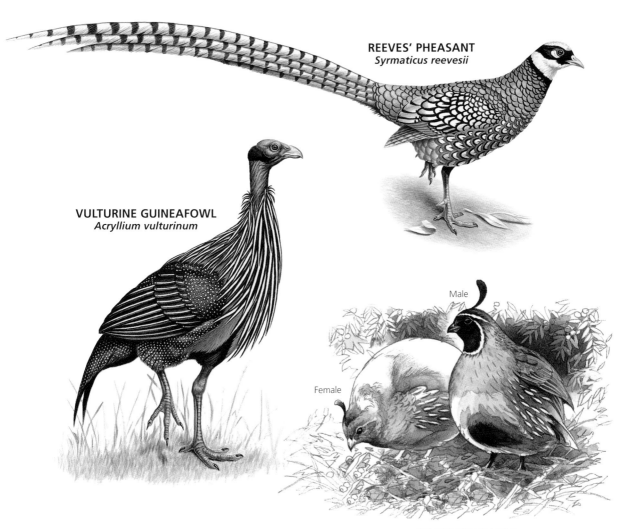

REEVES' PHEASANT
Syrmaticus reevesii

VULTURINE GUINEAFOWL
Acryllium vulturinum

Male

Female

GAMBEL'S QUAILS
Callipepla gambelii

SPRUCE GROUSE
Falcipennis canadensis

RUFFED GROUSE
Bonasa umbellus

277

Cranes

Location

■ Cranes

LIMPKIN
Aramus guarauna

RED-NECKED CRAKE
Rallina tricolor

WOODFORD'S RAIL
Nesoclopeus woodfordi

SIBERIAN CRANE MATING DANCE

Male initiates mating dance.   He performs rapid bows before stretching his wings and calling.   The female responds, and is imitated.

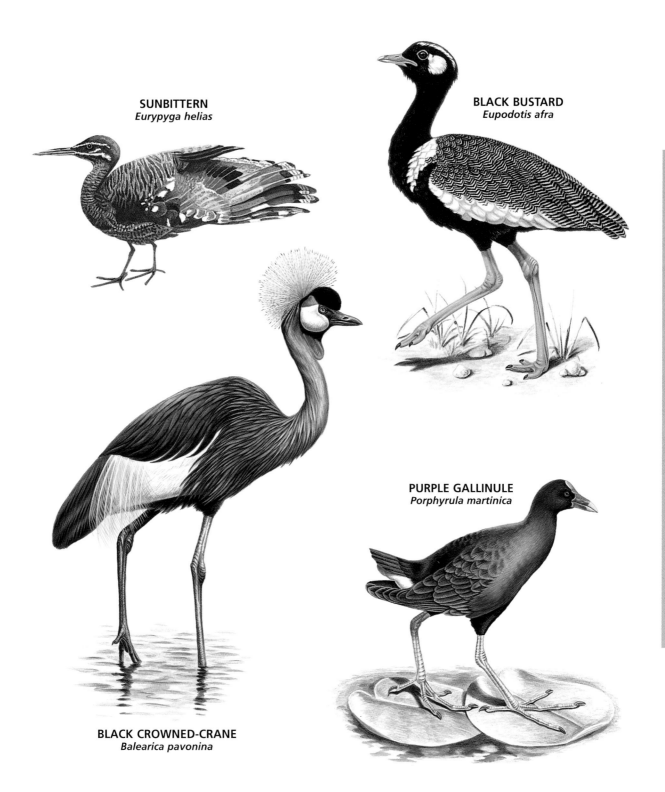

SUNBITTERN
Eurypyga helias

BLACK BUSTARD
Eupodotis afra

PURPLE GALLINULE
Porphyrula martinica

BLACK CROWNED-CRANE
Balearica pavonina

279

Cranes

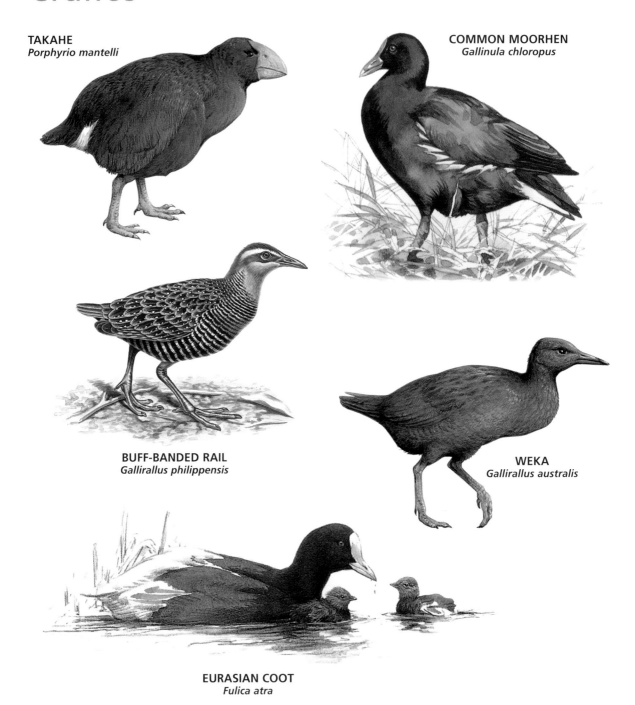

TAKAHE
Porphyrio mantelli

COMMON MOORHEN
Gallinula chloropus

BUFF-BANDED RAIL
Gallirallus philippensis

WEKA
Gallirallus australis

EURASIAN COOT
Fulica atra

Juvenile (left) and adult moorhens

Kagu

Sandhill cranes migrating

Water rail chicks

Water rail

Juvenile rail skeleton

Red-crowned crane

White-quilled bustard

Plains-wanderer

Okinawa rail

Common cranes displaying

Houbara bustard

Waders and Shorebirds

WADER AND SHOREBIRD BEHAVIOR

Oystercatchers
Individual oystercatchers specialize in catching specific foods, such as worms, crabs, oysters or mussels.

Sandwich terns
Male and female sandwich terns share small fishes as part of their courtship display during the breeding season.

Horned puffins feeding
Horned puffins dive deeply into the ocean to catch small fishes, which are held in their mouth by backward-facing grooves until they are ready to eat them.

Avocet with chicks
Avocets feed by sweeping their long, upturned bill through the mud while slowly walking forward.

Turnstones
As their name suggests, turnstones walk purposefully over the mudflats, turning over stones, seaweed or wood with their beaks in search of small crustaceans and worms.

Spotted sandpiper in summer plumage

Willets in winter (left) and breeding plumage

Far Eastern curlew bill

Arctic tern in flight

Jacana foot

Skua spreading wings

Godwit (top), dowitcher and curlew bills

Common murre egg

Golden plovers in flight

Black-headed gull facial plumage

Winter

Summer

Ruddy turnstone

Crab plovers digging nest holes

Waders

Location

■ Waders

COMMON SNIPE
Gallinago gallinago

BANDED LAPWING
Vanellus tricolor

BLACK-BELLIED PLOVER
Pluvialis squatarola

Summer plumage

Winter plumage

Winter plumage

SPOTTED SANDPIPER
Actitis macularia

KILLDEER
Charadrius vociferus
The killdeer acts out a 'broken wing' display to distract and draw predators from the nest.

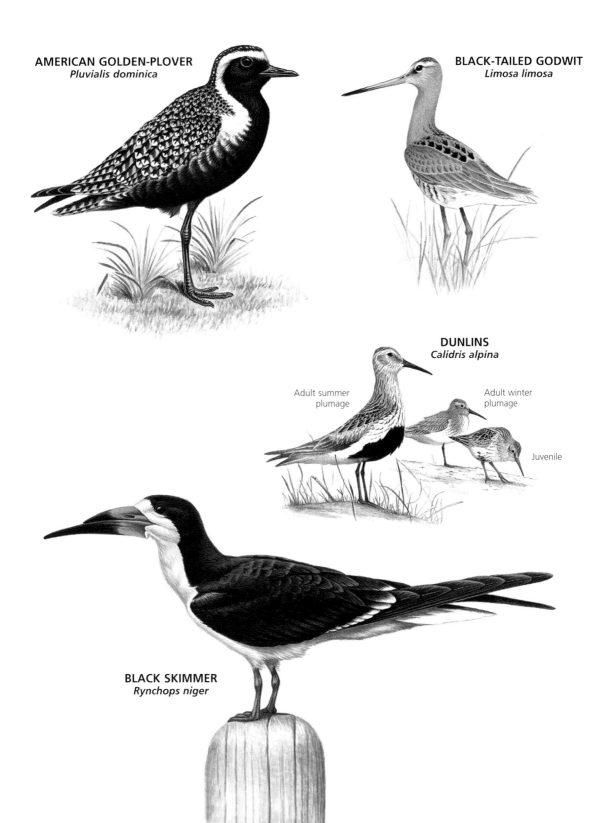

AMERICAN GOLDEN-PLOVER
Pluvialis dominica

BLACK-TAILED GODWIT
Limosa limosa

DUNLINS
Calidris alpina

Adult summer
plumage

Adult winter
plumage

Juvenile

BLACK SKIMMER
Rynchops niger

Shorebirds and Seabirds

Location

■ Shorebirds and seabirds

GULL WING CUTAWAY

Wing thicker at
front than back

SHOREBIRD HABITAT

Little tern

Northern
gannet

Pied avocet

Yellow-legged
gull

Cory's shearwater
(pelicaniform)

Great
cormorant

Little tern

Eurasian
curlew

Black-bellied
plover

RING-BILLED GULL
Larus delawarensis

ARCTIC TERN
Sterna paradisaea

CASPIAN TERNS
Terna caspia

Adult summer plumage

Juvenile

ATLANTIC PUFFIN
Fratercula arctica

YELLOW-LEGGED GULL
Larus cachinnans

MEW GULL
Larus canus

HERRING GULL
Larus argentatus

Juvenile herring gull

Common (left) and black guillemots

Common tern eggs hidden in rocks

Royal (top) and Caspian tern heads

287

Pigeons and Sandgrouse

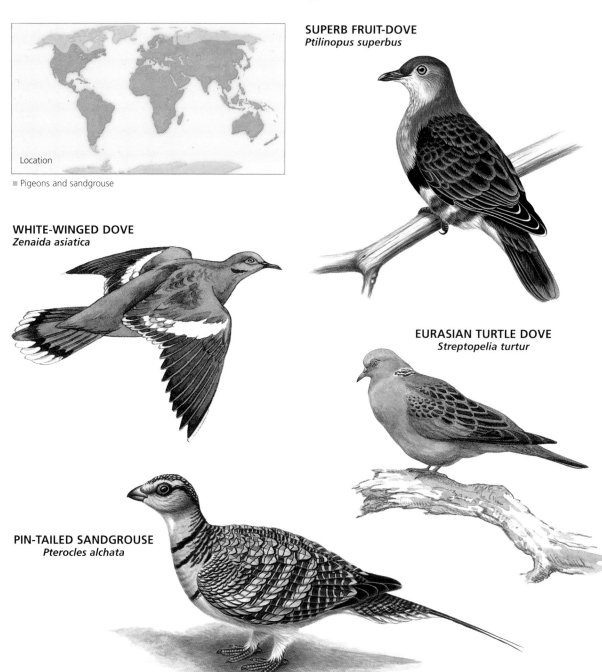

Location

■ Pigeons and sandgrouse

SUPERB FRUIT-DOVE
Ptilinopus superbus

WHITE-WINGED DOVE
Zenaida asiatica

EURASIAN TURTLE DOVE
Streptopelia turtur

PIN-TAILED SANDGROUSE
Pterocles alchata

288

ROCK DOVES
Columba livia

NICOBAR PIGEON
Caloenas nicobarica

Lichtenstein's sandgrouse

Male and female pin-tailed sandgrouse

Rock dove face and bill

Pigeon in flight

Crowned pigeon

White-winged dove nest

Parrots

PARROT SKULL

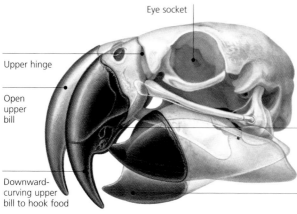

Eye socket

Upper hinge

Open upper bill

Downward-curving upper bill to hook food

Base of upper bill used to crack seeds against curve of lower bill

Open lower bill (mandible)

PALM COCKATOO HEAD AND BILL

The palm cockatoo uses its broad, curved bill to crush fruit, berries and seeds.

TYPES OF PARROT FEATHERS

Macaw body feather

Red-fan parrot facial feathers

Cockatiel flight feather

SCARLET MACAWS IN FLIGHT

Scarlet macaws (*Ara macao*) live in the rainforest canopies of Central and South America. Their brilliant, primary-colored plumage has made them popular with poachers and their numbers are now seriously depleted.

Australian night parrot

Kakapo

Black cockatoo

Kea

Scarlet macaw

Sulfur-crested cockatoo

Eastern rosella in flight

Gray parrot

Black-capped lory drinking nectar

Budgerigars

Lorikeet foot

Rosy-faced lovebird on branch

Parrots

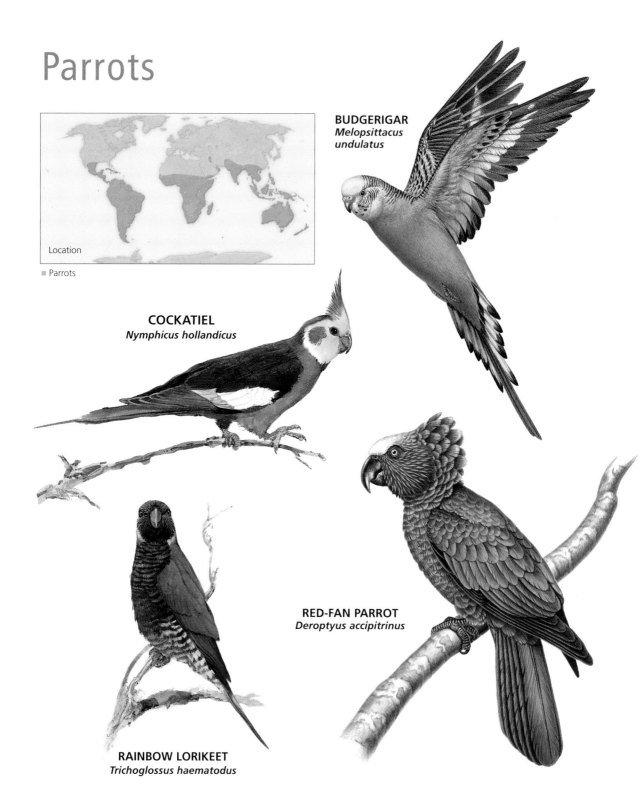

Location

■ Parrots

BUDGERIGAR
Melopsittacus undulatus

COCKATIEL
Nymphicus hollandicus

RED-FAN PARROT
Deroptyus accipitrinus

RAINBOW LORIKEET
Trichoglossus haematodus

BLACK-CAPPED LORY
Lorius lory

EASTERN ROSELLA
Platycercus eximius

PESQUET'S PARROT
Psittrichas fulgidus

SALMON-CRESTED COCKATOO
Cacatua moluccensis

Cuckoos and Turacos

Location

■ Cuckoos and turacos

WHITE-CHEEKED TURACO
Tauraco leucotis

GREAT BLUE TURACO
Corythaeola cristata

GREEN-BILLED MALKOHA
Phaenicophaeus tristis

**AFRICAN EMERALD
CUCKOO**
Chrysococcyx cupreus

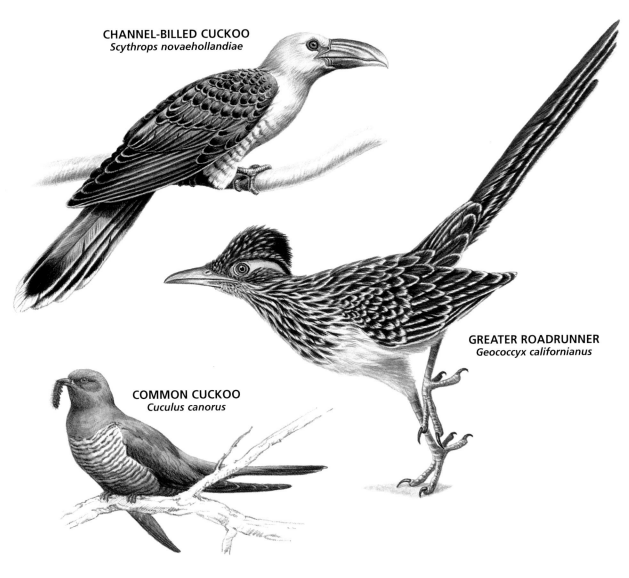

CHANNEL-BILLED CUCKOO
Scythrops novaehollandiae

GREATER ROADRUNNER
Geococcyx californianus

COMMON CUCKOO
Cuculus canorus

Cuckoo nest

Greater roadrunner
in defensive posture

Violet turaco on branch

295

Owls

Location

■ Owls

OWL HUNTING

Glides silently
toward prey

Raises wings
and extends
talons

Grabs prey with
talons, dropping
wings for balance

OWL SKULL

Some owl species have asymmetrical ears
to provide greater accuracy in locating prey.

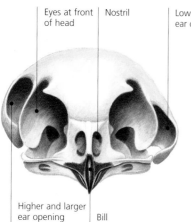

Eyes at front
of head

Nostril

Lower and smaller
ear opening

Higher and larger
ear opening

Bill

OWL VISION

70° binocular
vision range

110° field of vision range

Barn owl talons

Pectinate toenail
for preening

Opposable digits
for gripping

Flight feathers are serrated to reduce
noise.

Barn owl head and shoulders

SPOTTED OWL
Strix occidentalis

BARN OWL
Tyto alba

LITTLE OWL
Athene noctua

**GREAT HORNED OWL
WITH YOUNG**
Bubo virginianus

Snowy owl

Barn owl chicks

Barred owl chicks in nest

Nightjars and Frogmouths

Location

■ Nightjars and frogmouths

JAVAN FROGMOUTH
Batrachostomus javensis

SPOTTED NIGHTJAR
Eurostopodus argus

COMMON POTOO
Nyctibius griseus

TAWNY FROGMOUTH
Podargus strigoides

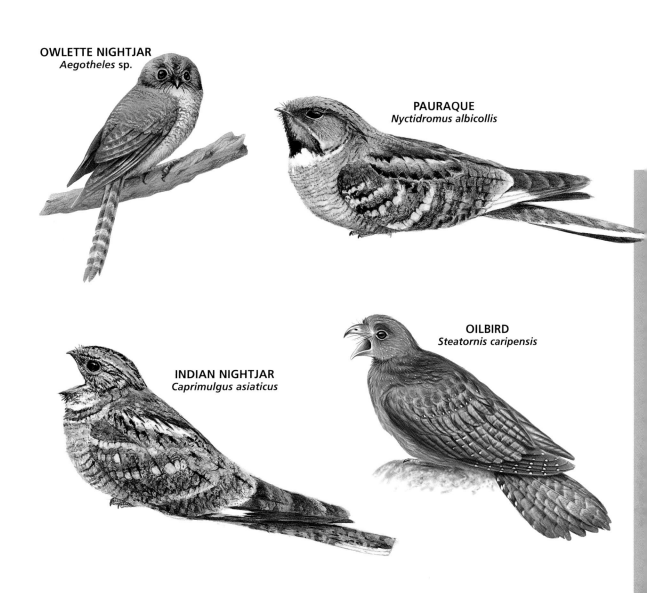

OWLLETTE NIGHTJAR
Aegotheles sp.

PAURAQUE
Nyctidromus albicollis

INDIAN NIGHTJAR
Caprimulgus asiaticus

OILBIRD
Steatornis caripensis

Oilbird showing back and tail plumage

Common nighthawk catching prey

Nightjar hunting

Hummingbirds and Swifts

PURPLE-BACKED THORNBILL HOVERING
The purple-backed thornbill's wings beat rapidly in
a number of directions, rotating around the flexible
shoulder joint to enable the bird to hover.

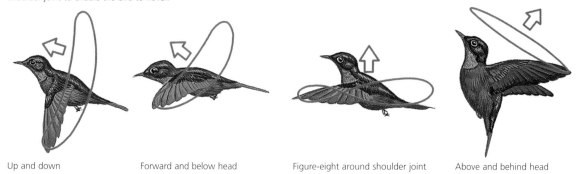

Up and down   Forward and below head   Figure-eight around shoulder joint   Above and behind head

HUMMINGBIRD FEEDING

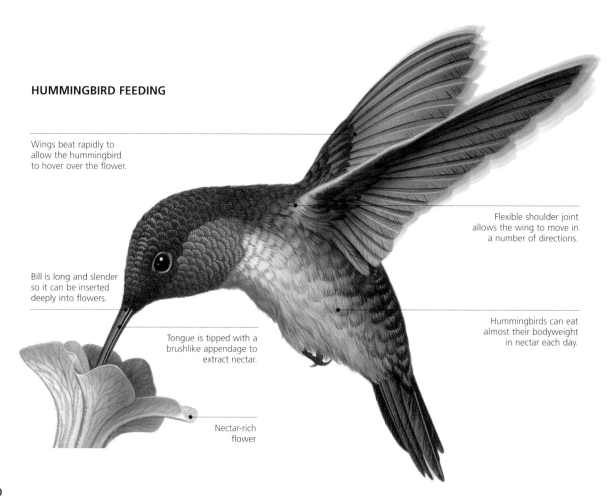

Wings beat rapidly to
allow the hummingbird
to hover over the flower.

Flexible shoulder joint
allows the wing to move in
a number of directions.

Bill is long and slender
so it can be inserted
deeply into flowers.

Hummingbirds can eat
almost their bodyweight
in nectar each day.

Tongue is tipped with a
brushlike appendage to
extract nectar.

Nectar-rich
flower

300

Hummingbird hovering

White-throated needletail gliding

Rufous hummingbird in flight

Costa's hummingbird feeding

Swifts in flight over rooftops

Black-chinned hummingbird head

Anna's hummingbird feeding

Anna's hummingbird egg

Black-chinned hummingbird hovering

Male ruby-throated hummingbird head

Ruby-throated hummingbird on nest

Chimney (top) and Vaux's swifts in flight

Hummingbirds and Swifts

Location

■ Hummingbirds and swifts

LITTLE SWIFT
Apus affinis

RUFOUS HUMMINGBIRD
Selasphorus rufus

**BLACK-CHINNED
HUMMINGBIRD**
Archilochus alexandri

RUBY-THROATED HUMMINGBIRD
Archilochus colubris

GRAY-RUMPED TREESWIFT
Hemiprocne longipennis

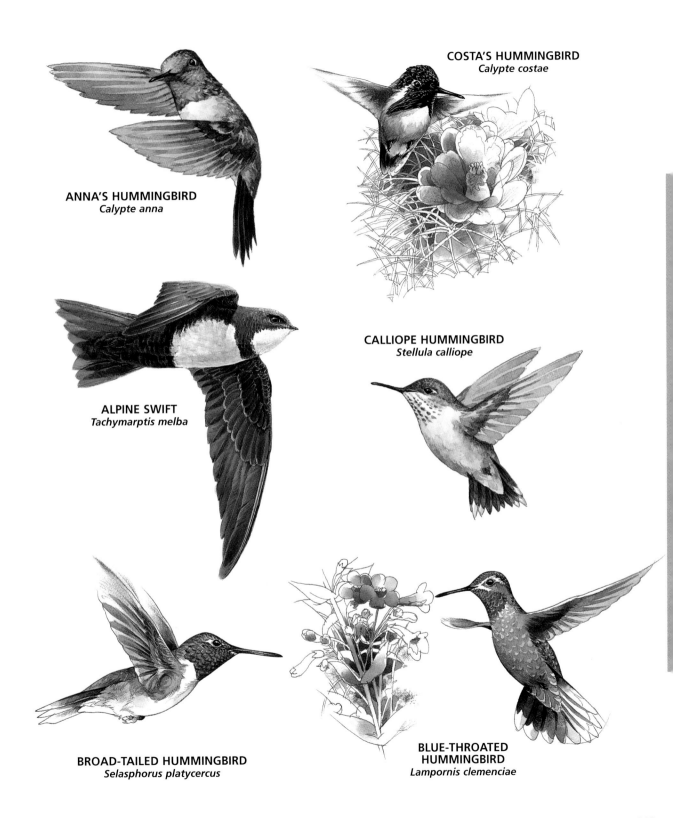

ANNA'S HUMMINGBIRD
Calypte anna

COSTA'S HUMMINGBIRD
Calypte costae

ALPINE SWIFT
Tachymarptis melba

CALLIOPE HUMMINGBIRD
Stellula calliope

BROAD-TAILED HUMMINGBIRD
Selasphorus platycercus

**BLUE-THROATED
HUMMINGBIRD**
Lampornis clemenciae

Mousebirds and Trogons

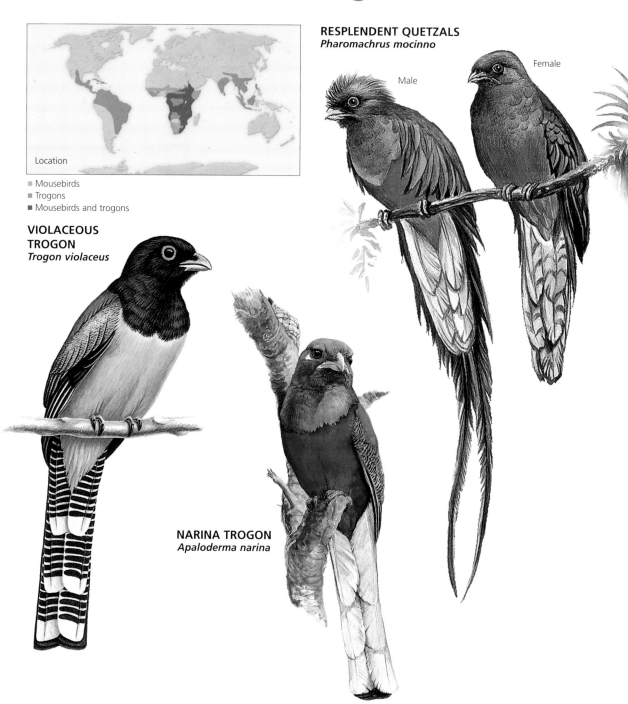

Location

- Mousebirds
- Trogons
- Mousebirds and trogons

RESPLENDENT QUETZALS
Pharomachrus mocinno

Male

Female

**VIOLACEOUS
TROGON**
Trogon violaceus

NARINA TROGON
Apaloderma narina

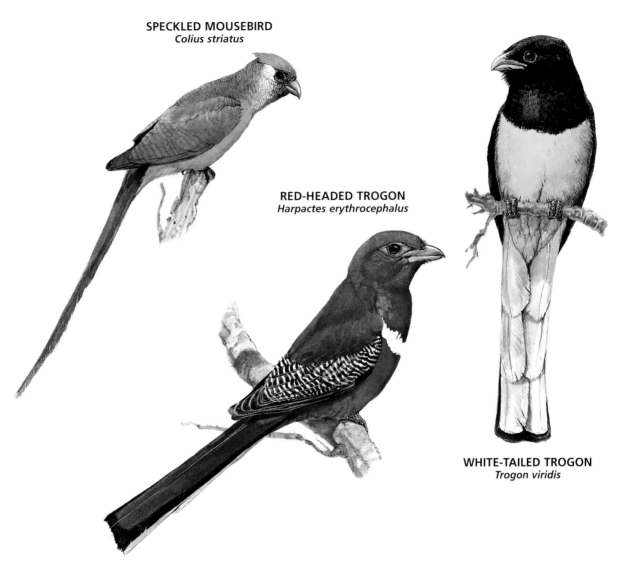

SPECKLED MOUSEBIRD
Colius striatus

RED-HEADED TROGON
Harpactes erythrocephalus

WHITE-TAILED TROGON
Trogon viridis

Scarlet-rumped trogon

Diard's trogon

Resplendent quetzal tail feathers

Kingfishers

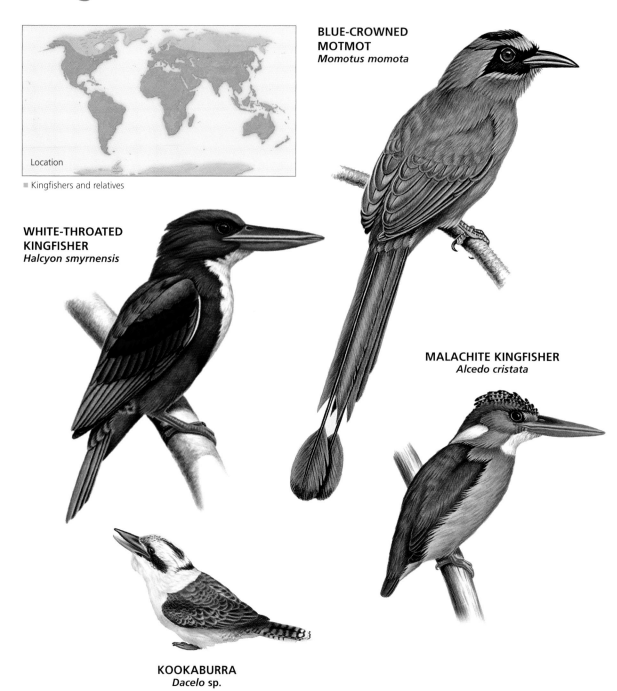

Location

■ Kingfishers and relatives

BLUE-CROWNED MOTMOT
Momotus momota

WHITE-THROATED KINGFISHER
Halcyon smyrnensis

MALACHITE KINGFISHER
Alcedo cristata

KOOKABURRA
Dacelo sp.

306

Cuban tody on branch

Rufous motmot

Eurasian kingfisher searching for prey

Banded kookaburra

Bee-eater in excavated nest cavity

Long-tailed ground-roller

Kookaburras on branch

Colorful kingfisher

Rufous-headed ground-roller

Hoopoe with young hidden in nest

Hoopoe with crest flattened

Hoopoe with crest erect

Kingfishers

**NORTHERN CARMINE
BEE-EATER**
Merops nubicus

**RAINBOW
BEE-EATER**
Merops ornatus

**BELTED KINGFISHER
FISHING**

DOLLARBIRD
Eurystomus orientalis

Female flies in search
of prey.

She spots a fish.

She swoops down and
dives into the water.

Catching the fish in her bill,
she returns to the nest.

The nest is a tunnel dug into a riverbank,
above the waterline to avoid flooding.

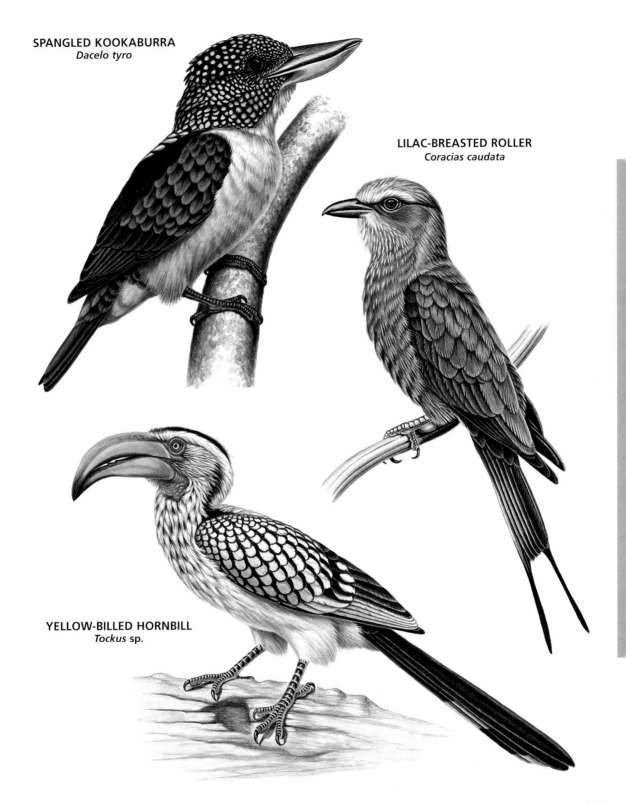

SPANGLED KOOKABURRA
Dacelo tyro

LILAC-BREASTED ROLLER
Coracias caudata

YELLOW-BILLED HORNBILL
Tockus sp.

Woodpeckers and Toucans

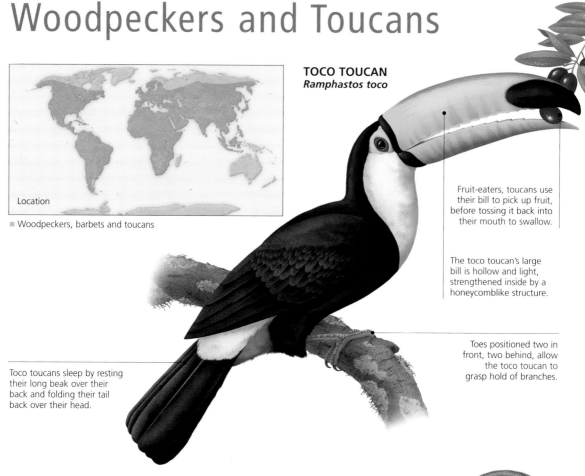

Location

■ Woodpeckers, barbets and toucans

TOCO TOUCAN
Ramphastos toco

Fruit-eaters, toucans use their bill to pick up fruit, before tossing it back into their mouth to swallow.

The toco toucan's large bill is hollow and light, strengthened inside by a honeycomblike structure.

Toes positioned two in front, two behind, allow the toco toucan to grasp hold of branches.

Toco toucans sleep by resting their long beak over their back and folding their tail back over their head.

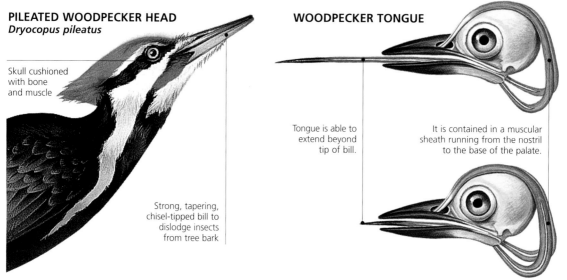

PILEATED WOODPECKER HEAD
Dryocopus pileatus

Skull cushioned with bone and muscle

Strong, tapering, chisel-tipped bill to dislodge insects from tree bark

WOODPECKER TONGUE

Tongue is able to extend beyond tip of bill.

It is contained in a muscular sheath running from the nostril to the base of the palate.

Great (left)
and lesser spotted woodpeckers

Maroon woodpecker on branch

Chestnut-capped puffbird

Yellow-rumped honeyguide

Green woodpecker digging in anthill

Red-bellied woodpecker

Acorn woodpecker storing acorn

Rufous piculet

Hairy woodpecker tapping tree bark

Acorn woodpecker in flight

Downy woodpecker

Hairy and downy woodpecker tails

311

NORTHERN FLICKER SUBSPECIES

Gilded flicker

Red-shafted flicker

Yellow-shafted flicker

NORTHERN FLICKER
Colaptes auratus

KEEL-BILLED TOUCAN
Ramphastos sulfuratos

RED-AND-YELLOW BARBET
Trachyphonus erythrocephalus

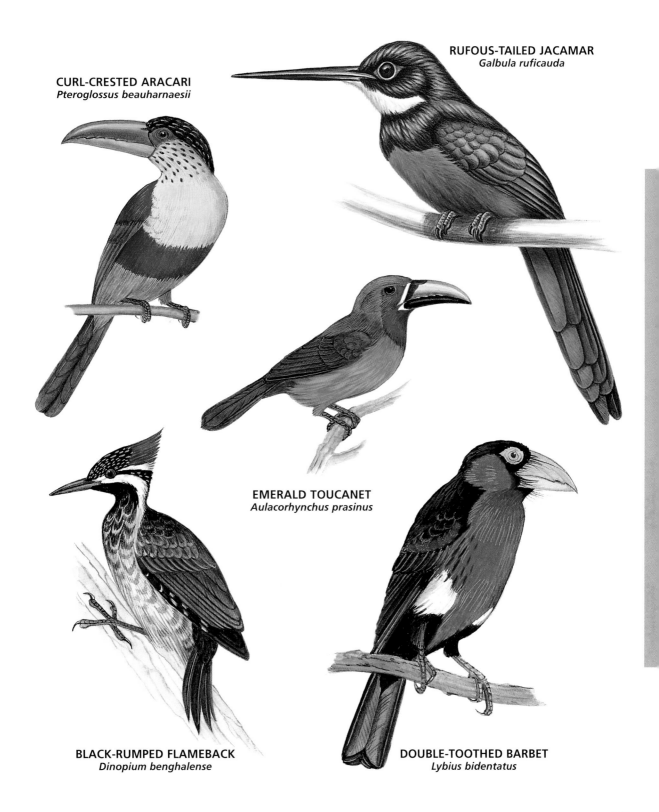

CURL-CRESTED ARACARI
Pteroglossus beauharnaesii

RUFOUS-TAILED JACAMAR
Galbula ruficauda

EMERALD TOUCANET
Aulacorhynchus prasinus

BLACK-RUMPED FLAMEBACK
Dinopium benghalense

DOUBLE-TOOTHED BARBET
Lybius bidentatus

Passerines

EVOLUTION OF PASSERINES

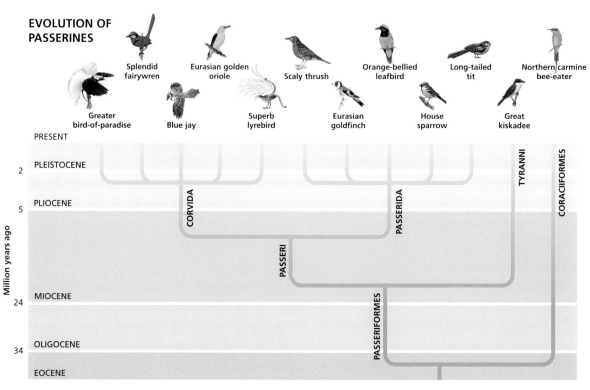

Splendid fairywren

Eurasian golden oriole

Scaly thrush

Orange-bellied leafbird

Long-tailed tit

Northern carmine bee-eater

Greater bird-of-paradise

Blue jay

Superb lyrebird

Eurasian goldfinch

House sparrow

Great kiskadee

Million years ago

PRESENT

PLEISTOCENE

2

PLIOCENE

5

MIOCENE

24

OLIGOCENE

34

EOCENE

CORVIDA

PASSERI

PASSERIDA

TYRANNI

CORACIIFORMES

PASSERIFORMES

TYPES OF PASSERINES

Passerines, or perching birds, are by far the most numerous order of birds, consisting of approximately 5,300 species, or half of all bird species.

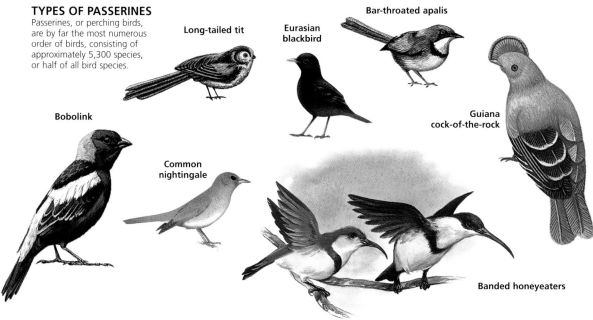

Bar-throated apalis

Long-tailed tit

Eurasian blackbird

Bobolink

Guiana cock-of-the-rock

Common nightingale

Banded honeyeaters

FLIGHT STAGES OF THE EUROPEAN ROBIN

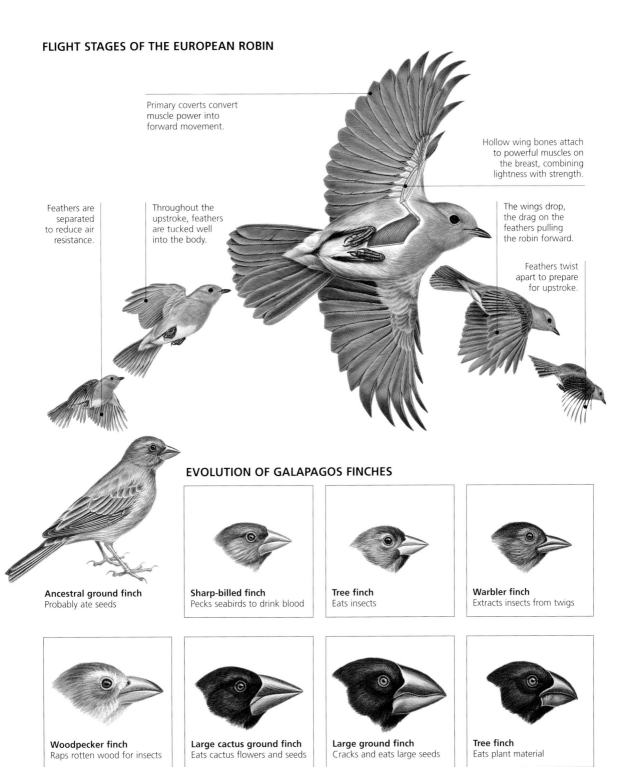

Primary coverts convert muscle power into forward movement.

Hollow wing bones attach to powerful muscles on the breast, combining lightness with strength.

Feathers are separated to reduce air resistance.

Throughout the upstroke, feathers are tucked well into the body.

The wings drop, the drag on the feathers pulling the robin forward.

Feathers twist apart to prepare for upstroke.

EVOLUTION OF GALAPAGOS FINCHES

Ancestral ground finch
Probably ate seeds

Sharp-billed finch
Pecks seabirds to drink blood

Tree finch
Eats insects

Warbler finch
Extracts insects from twigs

Woodpecker finch
Raps rotten wood for insects

Large cactus ground finch
Eats cactus flowers and seeds

Large ground finch
Cracks and eats large seeds

Tree finch
Eats plant material

Broadbills, Pittas, Ovenbirds and Tyrant Flycatchers

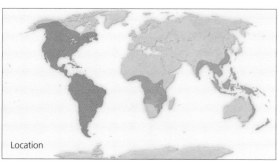

Location

- Broadbills and pittas
- Ovenbirds and tyrant flycatchers

GREEN BROADBILL
Calyptomena viridis

CASSIN'S KINGBIRD
Tyrannus vociferans

BANDED PITTA
Pitta guajana

GREAT KISKADEE
Pitangus sulphuratus

BLACK PHOEBE
Sayornis nigricans

GUIANA COCK-OF-THE-ROCK
Rupicola rupicola

EASTERN KINGBIRD
Tyrannus tyrannus

MALE WIRE-TAILED MANAKIN
Pipra filicauda

Rufous hornero

Western kingbird

Wire-tailed manakin skeleton

Lyrebirds, Scrub-birds, Larks, Wagtails and Swallows

Location

- Larks, wagtails and swallows
- Larks, wagtails, swallows, lyrebirds and scrub-birds

WAGTAILS
Motacilla sp.

Adult pied wagtail

Juvenile white wagtail   Adult white wagtail

WOODLARK
Lullula arborea

GRAY WAGTAIL
Motacilla cinerea

SKYLARKS FEEDING
Alauda arvensis

WATER PIPIT
Anthus spinoletta

Male (left) and female purple martins

Barn swallow

Male (left) and female swallows

Sand martins at nest cavity

Barn swallow tail feathers

Meadow (top) and rock pipits

Cliff swallow leaving nest

Prairie (left) and northern horned larks

Superb lyrebird displaying tail plumage

Cliff swallow tail feathers

British yellow wagtail head

Continental yellow wagtail head

Cuckooshrikes, Bulbuls, Leafbirds, Shrikes, Vangas and Waxwings

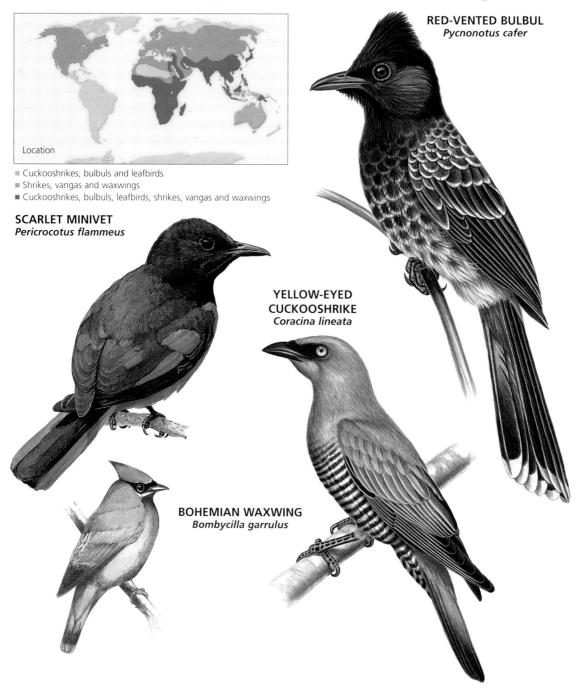

RED-VENTED BULBUL
Pycnonotus cafer

Location

- Cuckooshrikes, bulbuls and leafbirds
- Shrikes, vangas and waxwings
- Cuckooshrikes, bulbuls, leafbirds, shrikes, vangas and waxwings

SCARLET MINIVET
Pericrocotus flammeus

**YELLOW-EYED
CUCKOOSHRIKE**
Coracina lineata

BOHEMIAN WAXWING
Bombycilla garrulus

SCARLET MINIVET IN FLIGHT
Scarlet minivets fold their wings briefly after completing a
strong downstroke, to save energy before the next upstroke.

LONG-TAILED SHRIKE
Lanius schach

**GOLDEN-FRONTED
LEAFBIRD**
Chloropsis aurifrons

**WHITE-CHEEKED
BULBUL**
Pycnonotus leucogenys

**ORANGE-BELLIED
LEAFBIRD**
Chloropsis hardwickii

Mockingbirds, Accentors, Dippers, Thrushes, Babblers and Wrens

Location

- Thrushes, dippers, babblers and wrens
- Mockingbirds, thrushes, dippers, babblers and wrens
- Accentors, thrushes, dippers, babblers and wrens

SCALY THRUSH
Zoothera dauma

RED-BILLED LEIOTHRIX
Leiothrix lutea

WHITE-RUMPED SHAMA
Copsychus malabaricus

AMERICAN ROBIN FEEDING FLEDGLINGS
Turdus migratorius

Male American redstart on branch

Wren singing on branch

Wren at entrance of dome-shaped nest

Male black redstart

Canyon wren on rock

Male European robin 'redbreast'

Stonechats and whinchat (bottom)

Mockingbird in flight

Eurasian blackbird

Male western bluebird

Eastern bluebird

Dunnock

Warblers, Flycatchers, Fairywrens, Logrunners and Monarchs

Location

- Warblers and flycatchers
- Fairywrens, logrunners and monarchs
- Warblers, flycatchers, fairywrens, logrunners and monarchs

RED-CAPPED ROBIN
Petroica goodenovii

CRESTED SHRIKE-TIT
Falcunculus frontatus

GOLDCREST
Regulus regulus

FIRECREST
Regulus ignicapillus

LOGRUNNER
Orthonyx temminckii

GARDEN WARBLER
Sylvia borin

RUFOUS-BELLIED NILTAVA
Niltava sundara

SPECTACLED MONARCH
Monarcha trivirgatus

GOLDEN-CROWNED KINGLET
Regulus satrapa

SPLENDID FAIRYWREN
Malurus splendens

Tits, Nuthatches, Treecreepers and Honeyeaters

Location

- Tits
- Nuthatches, treecreepers and honeyeaters
- Tits, nuthatches, treecreepers and honeyeaters

GREAT TIT
Parus major

COAL TIT
Periparus ater

EASTERN SPINEBILL
Acanthorhynchus tenuirostris

MALE BLUE TIT
Cyanistes caeruleus

REGENT HONEYEATER
Xanthomyza phrygia

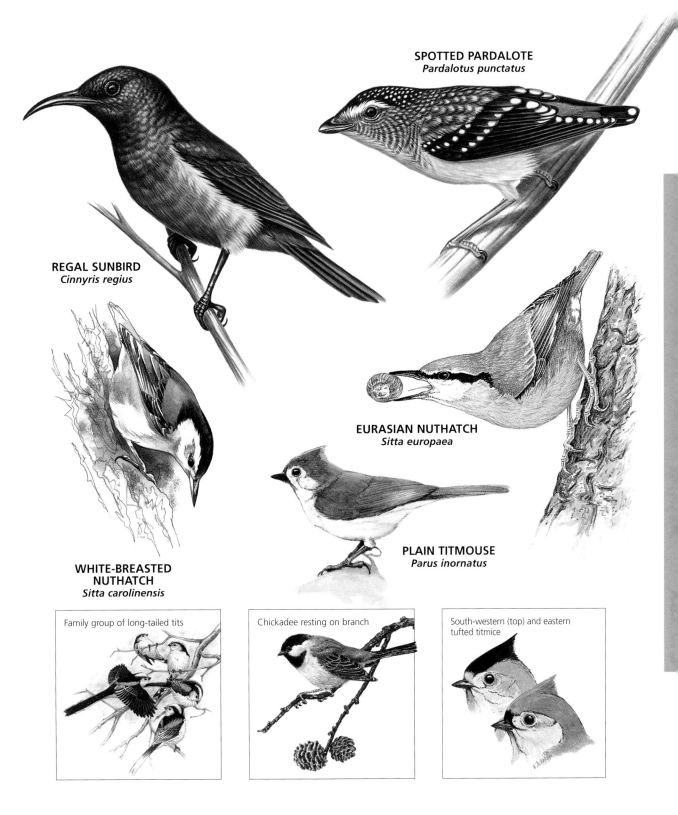

SPOTTED PARDALOTE
Pardalotus punctatus

REGAL SUNBIRD
Cinnyris regius

EURASIAN NUTHATCH
Sitta europaea

WHITE-BREASTED NUTHATCH
Sitta carolinensis

PLAIN TITMOUSE
Parus inornatus

Family group of long-tailed tits

Chickadee resting on branch

South-western (top) and eastern tufted titmice

327

Buntings, Tanagers, Vireos and Wood Warblers

Location

- Buntings
- Buntings, tanagers, vireos and wood warblers

FEMALE CHESTNUT-SIDED WARBLER
Dendroica pensylvanica

Breeding plumage

RED-LEGGED HONEYCREEPER
Cyanerpes cyaneus

JUVENILE WHITE-THROATED SPARROW
Zonotrichia albicollis

SEVEN-COLORED TANAGER
Tangara fastuosa

Female

Male in molt

Male with winter plumage

SCARLET TANAGERS
Piranga olivacea

PAINTED BUNTING
Passerina ciris

Passerines

BIRDS

Female northern cardinal head

Male northern cardinal

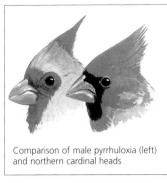

Comparison of male pyrrhuloxia (left) and northern cardinal heads

House sparrows

Slate-colored (top) and Oregon juncos

Scarlet tanager nest with eggs

Female house sparrow dustbathing

Northern parula at nest dug into moss

Adult white-striped sparrow head

Warbling (top) and Philadelphia vireos

Virginia's warbler on branch

Nashville warbler at nest entrance

Finches

Location

■ Finches

JAVA SPARROW
Padda oryzivora

**GREEN-WINGED
PYTILIA**
Pytilia melba

RED BISHOP
Euplectes orix

GOULDIAN FINCH
Chloebia gouldiae

Eurasian bullfinch at birth
Naked skin; eyes closed

At six days
Sparse covering of down; eyes open

At 28 days
Feathers grow out of tracts along wings.

At nine months
Bullfinch fully grown

EUROPEAN GOLDFINCH
Carduelis carduelis

EUROPEAN GREENFINCHES
Carduelis chloris

Male

Female

CHAFFINCH
Fringilla coelebs

Starlings, Magpie-larks and Icterids

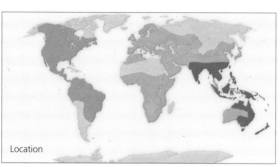

Location

- Starlings
- Magpie-larks (Old World) and icterids (New World)
- Starlings, magpie-larks and icterids

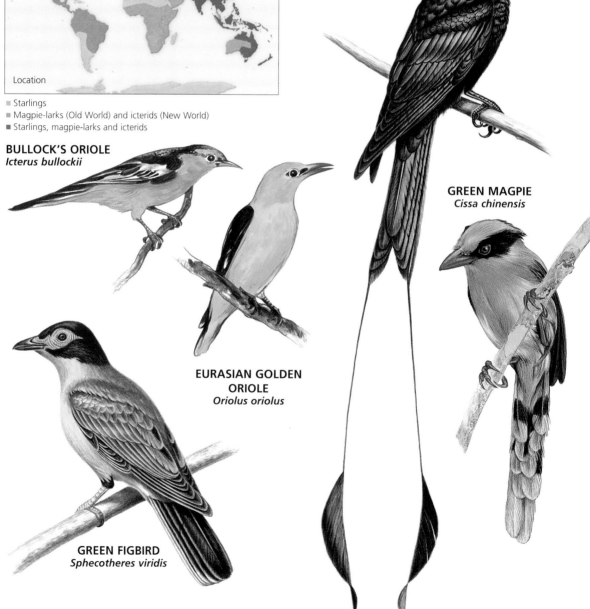

GREATER RACKET-TAILED DRONGO
Dicrurus paradiseus

BULLOCK'S ORIOLE
Icterus bullockii

GREEN MAGPIE
Cissa chinensis

EURASIAN GOLDEN ORIOLE
Oriolus oriolus

GREEN FIGBIRD
Sphecotheres viridis

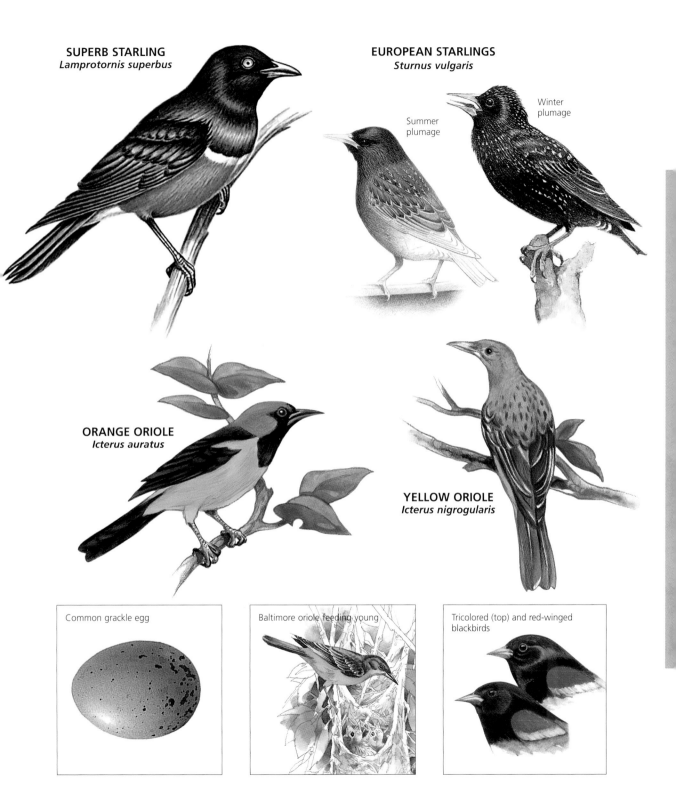

SUPERB STARLING
Lamprotornis superbus

EUROPEAN STARLINGS
Sturnus vulgaris

Summer
plumage

Winter
plumage

ORANGE ORIOLE
Icterus auratus

YELLOW ORIOLE
Icterus nigrogularis

Common grackle egg

Baltimore oriole feeding young

Tricolored (top) and red-winged
blackbirds

Bowerbirds and Birds-of-Paradise

Location

■ Bowerbirds and birds-of-paradise

BIRD-OF-PARADISE BEHAVIOR

Male birds-of-paradise show off their decorative plumage to attract females.

Females choose the male with the most brilliant plumage as their mate.

It was once thought that birds-of-paradise had no feet, because of incomplete specimens.

KING BIRD-OF-PARADISE
Cicinnurus regius

GREATER BIRD-OF-PARADISE
Paradisaea apoda

WESTERN PAROTIA
Parotia sefilata

GOLDEN BOWERBIRD
Pronodura newtoniana

MACGREGOR'S BOWERBIRD
Amblyornis macgregoriae

SATIN BOWERBIRD
Ptilonorhynchus violaceus

TOOTH-BILLED BOWERBIRD
Ailuroedus dentirostris

BOWER DECORATIONS

Male bowerbirds create elaborate bowers, or courtship display areas, in one of four different types—court, mat, maypole or avenue. These are decorated with a variety of found objects, from feathers to clothes-pegs.

Moss

Twigs

Feather

Crows and Jays

Location

■ Crows and jays

RED-BILLED CHOUGH
Pyrrhocorax pyrrhocorax

AMERICAN CROW FLIGHT
Corvus brachyrhynchos

Crows beat their wings evenly, with strong downstrokes, to maintain a level flight path.

YELLOW-BILLED CHOUGH
Pyrrhocorax graculus

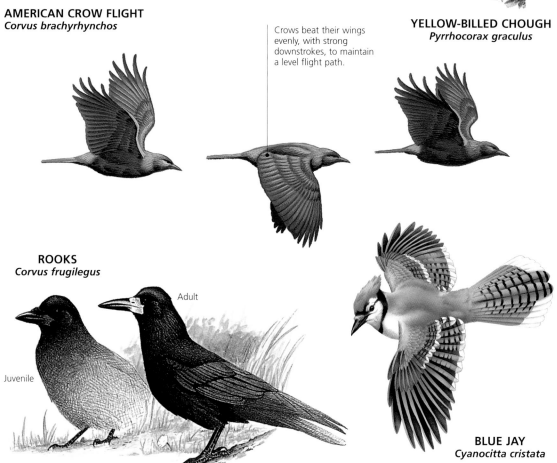

ROOKS
Corvus frugilegus

Adult

Juvenile

BLUE JAY
Cyanocitta cristata

FLORIDA SCRUB-JAY
Aphelocoma coerulescens

Scrub-jays are opportunistic hunters, often stealing eggs from other songbirds' nests.

EURASIAN JACKDAW
Corvus monedula

EURASIAN JAY
Garrulus glandarius

Yellow-billed magpie head

Blue jay head showing markings

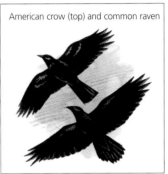

American crow (top) and common raven

Male pinyon jay on pine branch

Crested jay

Common raven head

REPTILES

Classifying Reptiles

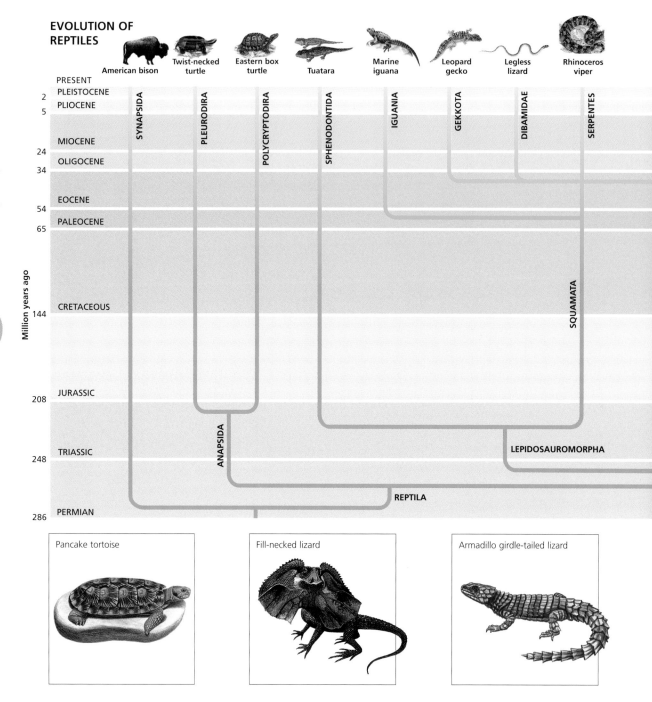

EVOLUTION OF REPTILES

American bison · Twist-necked turtle · Eastern box turtle · Tuatara · Marine iguana · Leopard gecko · Legless lizard · Rhinoceros viper

PRESENT
PLEISTOCENE
2
PLIOCENE
5

MIOCENE
24
OLIGOCENE
34

EOCENE
54
PALEOCENE
65

CRETACEOUS
144

JURASSIC
208

TRIASSIC
248

PERMIAN
286

Million years ago

SYNAPSIDA · PLEURODIRA · POLYCRYPTODIRA · SPHENODONTIDA · IGUANIA · GEKKOTA · DIBAMIDAE · SERPENTES

ANAPSIDA

SQUAMATA

LEPIDOSAUROMORPHA

REPTILA

Pancake tortoise

Fill-necked lizard

Armadillo girdle-tailed lizard

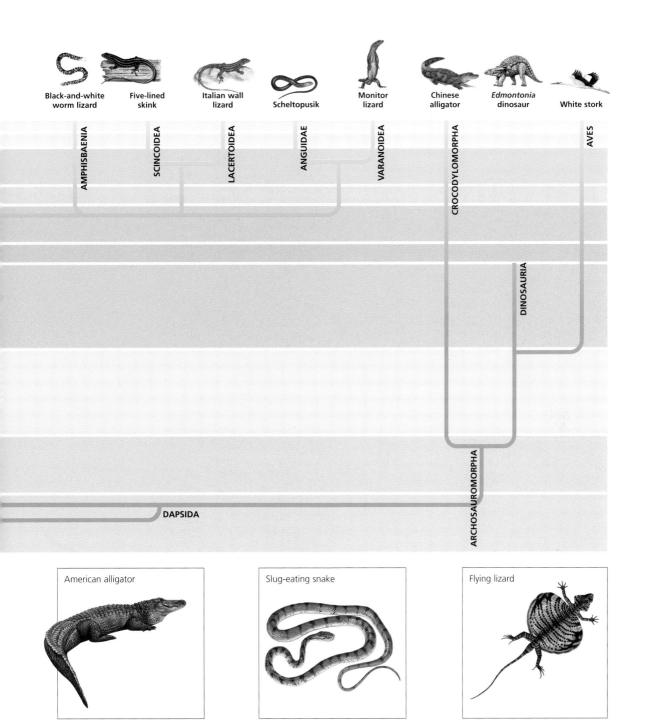

Black-and-white worm lizard | Five-lined skink | Italian wall lizard | Scheltopusik | Monitor lizard | Chinese alligator | *Edmontonia* dinosaur | White stork

AMPHISBAENIA

SCINCOIDEA

LACERTOIDEA

ANGUIDAE

VARANOIDEA

CROCODYLOMORPHA

AVES

DINOSAURIA

ARCHOSAUROMORPHA

DAPSIDA

American alligator

Slug-eating snake

Flying lizard

Reptile Characteristics

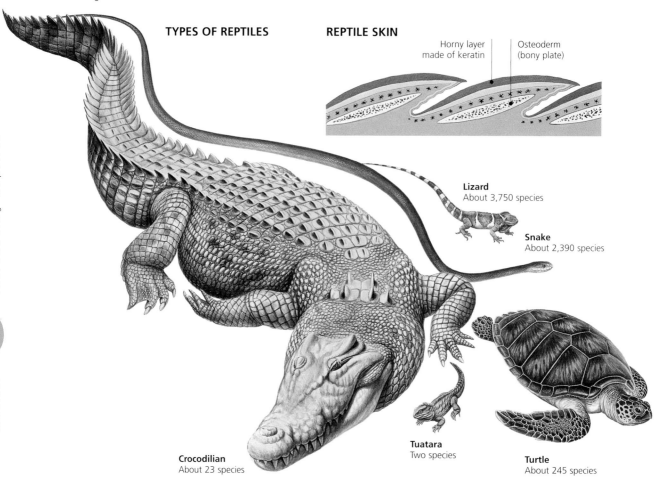

TYPES OF REPTILES

REPTILE SKIN

Horny layer made of keratin

Osteoderm (bony plate)

Lizard
About 3,750 species

Snake
About 2,390 species

Tuatara
Two species

Turtle
About 245 species

Crocodilian
About 23 species

Reptile Egg

Chorion

Embryo

Yolk sac

Allantonic (waste) cavity

REPTILE SCALES

Granular scales

Keeled scales

Smooth scales

REGULATING TEMPERATURE
Sand lizard

Noon
Shelters during the
heat of the day to
avoid overheating.

Early afternoon
Actively hunting,
mating and
defending territory.

Mid-morning
A warm body
provides energy
to be active.

Late afternoon
Basks and digests
the day's meals.

Early morning
Basks in the sun
to warm its body.

Night
Sleeps curled up
to maintain warmth.

Dawn
Wakes with the
sun and emerges
from shelter.

Sunset
The sand lizard
moves into shelter
before it cools down.

Evolution of Reptiles

EARLY REPTILES

Coelurosauravus
Late Permian

PTERANODON
With a wingspan of 24 ½ feet
(7.5 m), *Pteranodon* lived
85–75 million years ago.

Hylonomus
Carboniferous

Dimetrodon
Permian

Proganochelys
Late Triassic

DEINONYCHUS
Active dinosaur that lived
110–100 million years ago.

ICHTHYOSAURUS
Aquatic reptile that lived
206–150 million years ago.

Pteranodon
Late Cretaceous

Stegosaurus
Late Jurassic

Pachyrachis
Early Cretaceous

Planocephalosaurus
Late Triassic

Deinosuchus
Cretaceous

Archelon
Late Cretaceous

Ichthyosaurus
Early Jurassic

Elasmosaurus
Late Cretaceous

Turtles and Tortoises

LOGGERHEAD TURTLE 'CRYING'

Sea turtles, such as loggerhead turtles, 'weep' from glands around their eyes to remove salt from them. These tears are washed away by the ocean, so are visible only on land.

TORTOISE BURROW

Gopher tortoises dig burrows in the desert to escape the intensity of the heat of the day and chill of the night.

TURTLE REPRODUCTION

Mating
While some turtle species mate at sea, others mate on land. Almost all turtle species, however, lay their eggs on land.

Laying eggs
Most turtles and tortoises lay their eggs in a nest chamber, which keeps the developing eggs at a constant temperature.

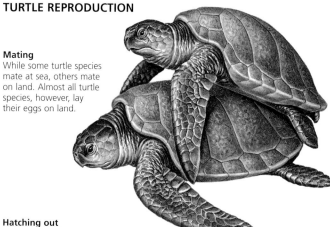

Hatching out
Newly hatched turtles dig themselves out of the nest chamber. They are independent from the moment they are born.

TUCKED AWAY

Most turtles are able to draw their necks and limbs completely into their shell, protecting themselves from predators.

FROM ABOVE AND BELOW

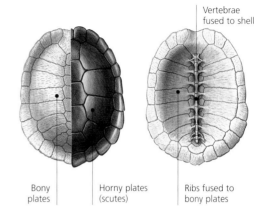

Vertebrae fused to shell

Bony plates

Horny plates (scutes)

Ribs fused to bony plates

TURTLE SHELLS

Land tortoise
Domed shell

Semi-terrestrial turtle
Flattened shell

Pond turtle
Small flattened shell

Sea turtle
Streamlined shell

TURTLE LIMBS

Land tortoise

Pond turtle

Sea turtle

Side-necked Turtles and Land Tortoises

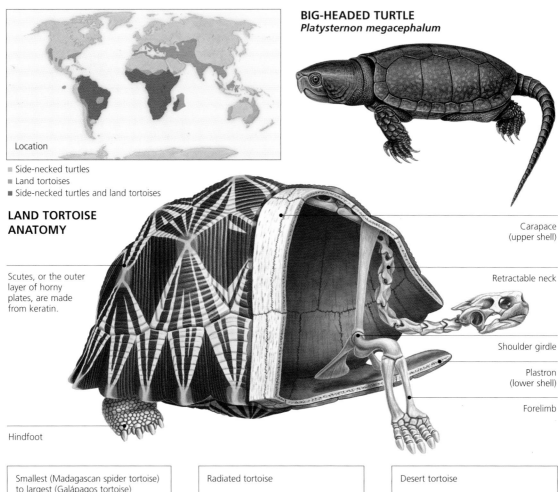

Location

- Side-necked turtles
- Land tortoises
- Side-necked turtles and land tortoises

BIG-HEADED TURTLE
Platysternon megacephalum

LAND TORTOISE ANATOMY

Scutes, or the outer layer of horny plates, are made from keratin.

Carapace
(upper shell)

Retractable neck

Shoulder girdle

Plastron
(lower shell)

Forelimb

Hindfoot

Smallest (Madagascan spider tortoise) to largest (Galápagos tortoise)

Radiated tortoise

Desert tortoise

YELLOW-SPOTTED AMAZON RIVER TURTLE
Podocnemis unifilis

TWIST-NECKED TURTLE
Platemys platycephala platycephala

SADDLEBACK TORTOISES
Geochelone sp.

Cactus plants form a major part of the saddleback tortoise's diet when there is no groundwater to drink.

The tortoises can stretch their necks out to reach high-growing food, or tuck them fully into their shells.

By lowering their hindlegs to the ground, saddleback tortoises can move their inflexible bodies higher.

The domed shells of many large tortoise species are honeycombed, not solid, to be less heavy.

Turtles

Location

■ Turtles

MALE PAINTED TURTLE
Chrysemys picta belli

MALAYAN SNAIL-EATING TURTLE
Malayemys subtrijuga

**SOUTHERN LOGGERHEAD
MUSK TURTLE**
Sternotherus minor minor

EASTERN BOX TURTLE
Terrapene carolina carolina

SPINED TURTLE
Heosemys spinosa

Raised nostrils on snout allow turtle to breathe out of water.

Forelimbs have evolved into flippers.

Marine turtles build up large fat reserves to feed off when migrating to breed and lay eggs.

PACIFIC HAWKSBILL TURTLE
Eretmochelys imbricata bissa

EASTERN SPINY SOFTSHELL TURTLE
Apalone spinifera spinifera

SOFTSHELL TURTLES
Unlike most turtle species, softshell turtles do not have hard carapaces, but fattened shells of leathery skin. Well-adapted for life in the water, they have webbed feet and a snorkle-shaped snout.

Crocodilians

CROCODILE HATCHING

GUARDING THE NEST
Female crocodilians cover their egg nests with warm rotting plant material, and do not leave until their young are hatched.

CARRYING YOUNG
Alligator hatchlings are carried in the mother's mouth to a nearby pond. The female will protect the hatchlings from predators for several weeks.

CROCODILIAN SKULLS

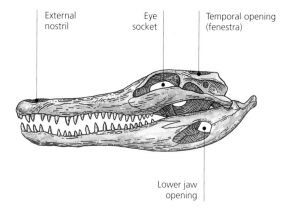

Temporal opening (fenestra)

External nostril

Eye socket

Temporal opening (fenestra)

External nostril

Eye socket

Lower jaw opening

354

CROCODILE GAITS

Crawling

Walking

Galloping

SNOUT COMPARISON

American crocodile

Black caiman

Gharial

ANKLE COMPARISON

Crocodilian   **Typical reptile**

Tibia

Fibula

Astragalus

Line of hinge

Tibia

Fibula

Astragalus

BREATHING HALF-SUBMERGED

External nostrils remain above water surface.

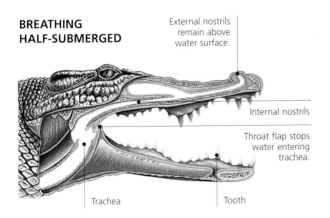

Internal nostrils

Throat flap stops water entering trachea.

Trachea   Tooth

CROCODILE TOOTH

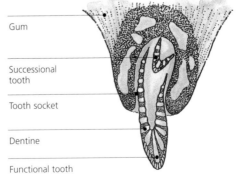

Gum

Successional tooth

Tooth socket

Dentine

Functional tooth

Alligators and Caimans

Location

- Alligators
- Caimans

COMMON CAIMAN
Caiman crocodilus

AMERICAN ALLIGATOR HUNTING
Alligator mississippiensis

American alligators often hunt near water-bird colonies, where they feed on the fish that gather to eat guano. The birds are not safe either—alligators can leap from the water to snatch fledglings or unwary adults from the branches.

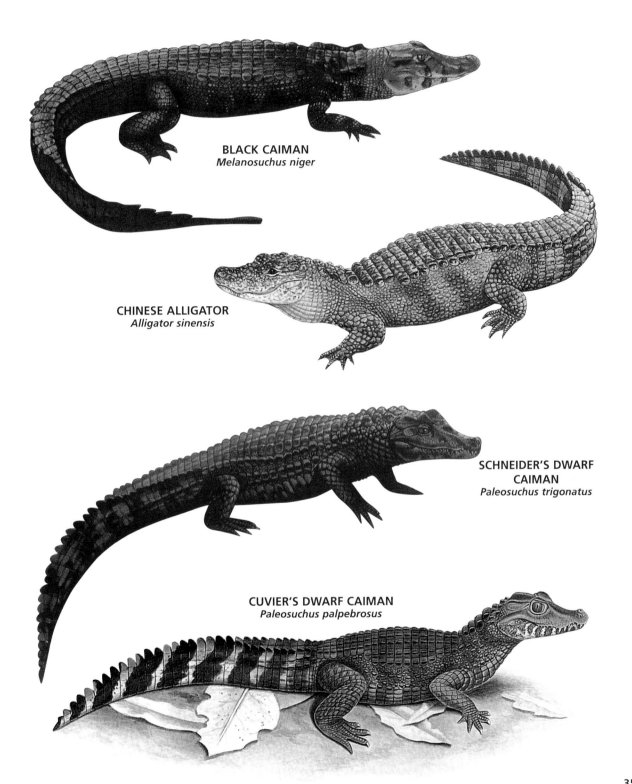

BLACK CAIMAN
Melanosuchus niger

CHINESE ALLIGATOR
Alligator sinensis

SCHNEIDER'S DWARF CAIMAN
Paleosuchus trigonatus

CUVIER'S DWARF CAIMAN
Paleosuchus palpebrosus

Crocodiles and Gharials

Location

- Crocodiles
- Gharials
- Crocodiles and gharials

SIAMESE CROCODILE
Crocodylus siamensis

CUBAN CROCODILE
Crocodylus rhombifer

**AFRICAN SLENDER-SNOUTED
CROCODILE**
Crocodylus cataphractus

GHARIAL
Gavialis gangeticus

MALAYAN GHARIAL
Tomistoma schlegelii

NILE CROCODILE
Crocodylus niloticus

ORINOCO CROCODILE
Crocodylus intermedius

CROCODILE SUBMERGED
Conservative hunters, crocodiles lie submerged
with only their eyes, ears and noses above
the water, waiting for prey. When they strike,
crocodiles move undetected through the water
and lunge up onto the bank, grabbing their
prey before quickly drowning, then eating it.

Amphisbaenians and Tuatara

Location

- Tuatara
- Amphisbaenians (worm-lizards)

ADAPTATIONS FOR DIGGING

Round-snouted
Push forward, then around

Spade-snouted
Push forward, then up

Chisel-snouted
Rotate in opposing directions

Keel-snouted
Push forward, then sideways

BLACK AND WHITE WORM-LIZARD
Amphisbaena fuliginosa

KEEL-SNOUTED WORM-LIZARD
Family *Amphisbaenidae*

SPADE-SNOUTED WORM-LIZARD
Family *Amphisbaenidae*

TUATARA

Tuatara can be found only on about 30 small islands off the northern coast of New Zealand.

Teeth fused to jawbone

'Tuatara' means 'lightning back' in Maori, which refers to the crest of spines along the males' back.

Tuatara are not fast runners, so lie in wait for prey to approach close enough to pounce.

TUATARA
Sphenodon sp.

Female

Male

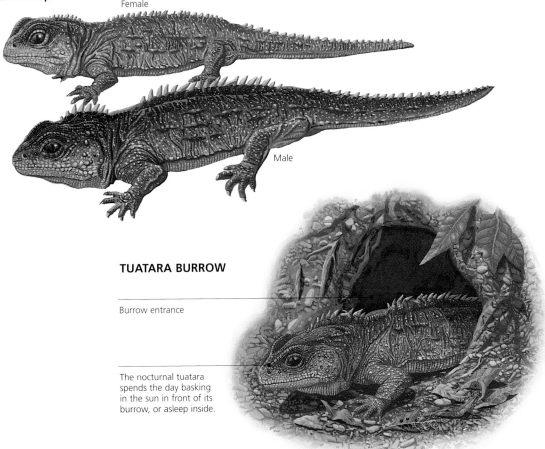

TUATARA BURROW

Burrow entrance

The nocturnal tuatara spends the day basking in the sun in front of its burrow, or asleep inside.

Lizards

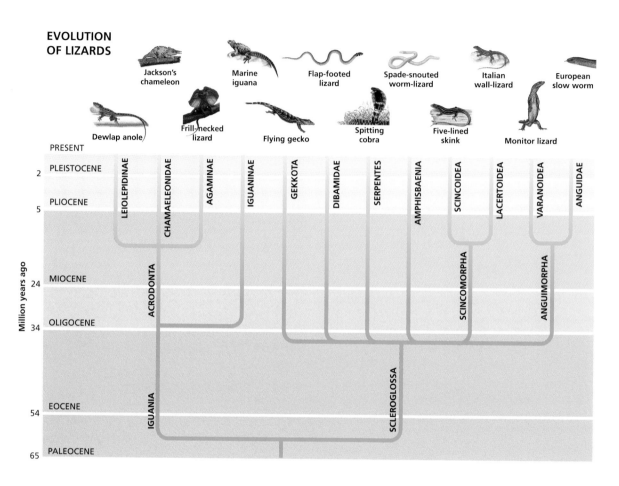

EVOLUTION OF LIZARDS

Jackson's chameleon

Marine iguana

Flap-footed lizard

Spade-snouted worm-lizard

Italian wall-lizard

European slow worm

Dewlap anole

Frill-necked lizard

Flying gecko

Spitting cobra

Five-lined skink

Monitor lizard

PRESENT

Million years ago	
2	PLEISTOCENE
5	PLIOCENE
24	MIOCENE
34	OLIGOCENE
54	EOCENE
65	PALEOCENE

LEIOLEPIDINAE

CHAMAELEONIDAE

AGAMINAE

IGUANINAE

GEKKOTA

DIBAMIDAE

SERPENTES

AMPHISBAENIA

SCINCOIDEA

LACERTOIDEA

VARANOIDEA

ANGUIDAE

ACRODONTA

IGUANIA

SCINCOMORPHA

ANGUIMORPHA

SCLEROGLOSSA

LIZARD TOES

Lizard toes have adapted to specific uses and environments, such as burrowing or climbing.

No toes

Reduced toes

Well-developed toes

FOOT SPECIALIZATIONS

Fringe-toed lizard

Fringe helps to cool feet.

Desert gecko

Webbing to grip on sand

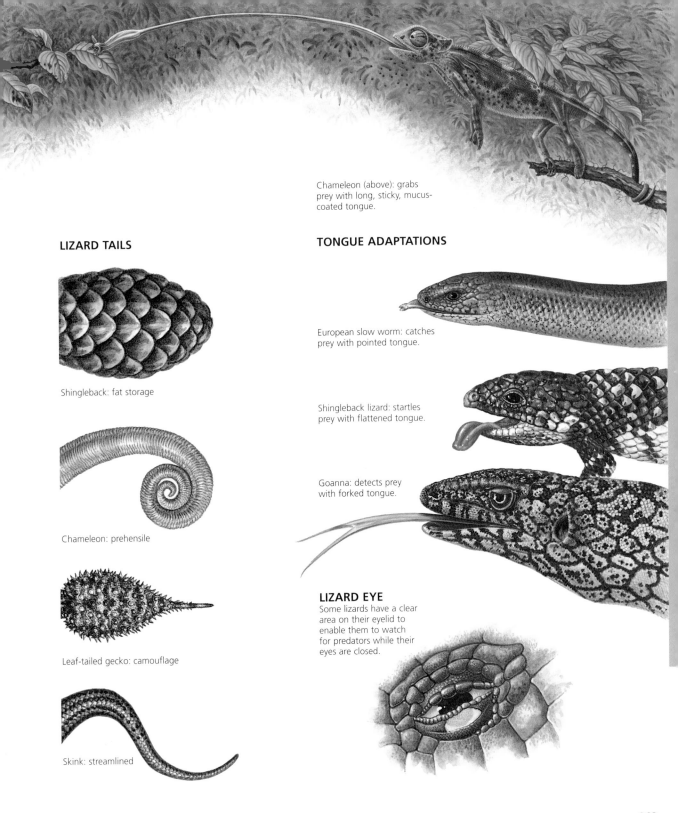

Chameleon (above): grabs prey with long, sticky, mucus-coated tongue.

LIZARD TAILS

Shingleback: fat storage

Chameleon: prehensile

Leaf-tailed gecko: camouflage

Skink: streamlined

TONGUE ADAPTATIONS

European slow worm: catches prey with pointed tongue.

Shingleback lizard: startles prey with flattened tongue.

Goanna: detects prey with forked tongue.

LIZARD EYE
Some lizards have a clear area on their eyelid to enable them to watch for predators while their eyes are closed.

Iguanids

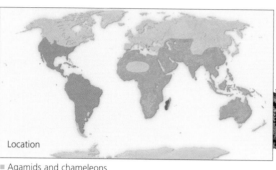

Location

- Agamids and chameleons
- Iguanas and anoles
- Agamids, chameleons, iguanas and anoles

HELMETED IGUANA HEAD

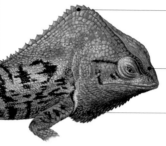

The function of the casque, or crest, extending from head to neck, is unknown.

Like chameleons, helmeted iguanas can change color.

Pouch under the chin expands when iguana defends its territory.

COLLARED LIZARD BEHAVIOR

Crouches down | Pushes up to appear bigger | Crouches down again

Rhinoceros iguanas change color with the heat of the sun.

Thorny devil

Basilisk running on water

Regal horned lizard squirting blood

Common iguana head

Bearded dragon expanding throat

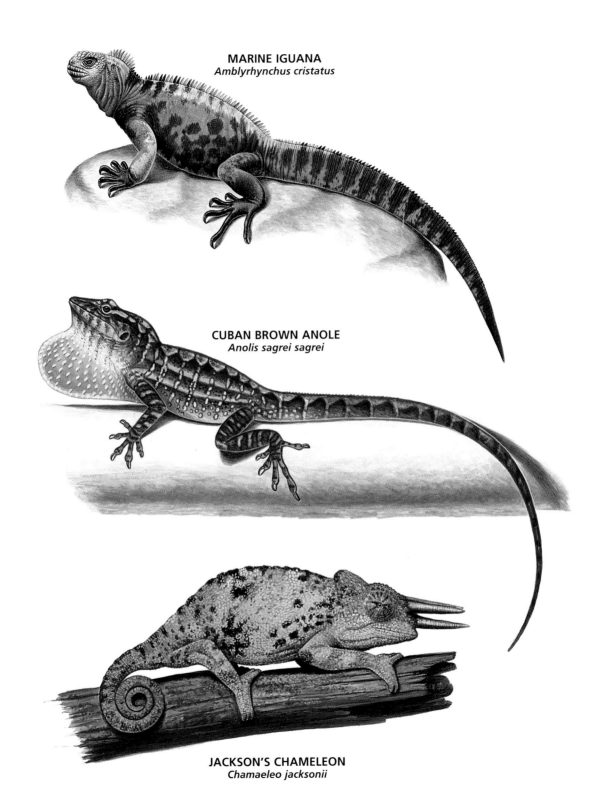

MARINE IGUANA
Amblyrhynchus cristatus

CUBAN BROWN ANOLE
Anolis sagrei sagrei

JACKSON'S CHAMELEON
Chamaeleo jacksonii

365

Agamids and Chameleons

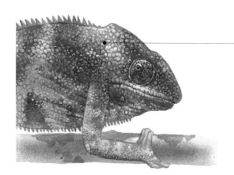

CHANGING COLOR

Male chameleons flush with color to warn rival males off their territories. This species changes from camouflaging green to an angry red.

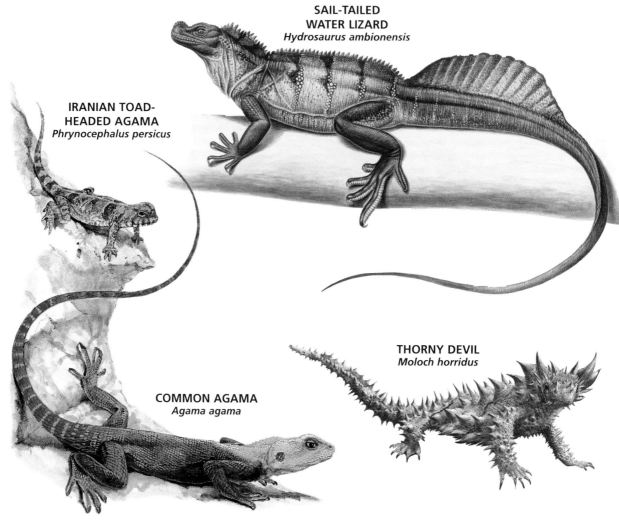

SAIL-TAILED WATER LIZARD
Hydrosaurus ambionensis

IRANIAN TOAD-HEADED AGAMA
Phrynocephalus persicus

COMMON AGAMA
Agama agama

THORNY DEVIL
Moloch horridus

MILITARY DRAGON
Ctenophorus isolepis gularis

TREE DRAGON
Diporiphora superba

FRILL-NECKED LIZARD
Chlamydosaurus kingii

Frill-necked lizards frighten off attackers by opening their mouth wide and hissing loudly.

Opening their frill rapidly, frill-necked lizards appear to increase in size.

MALAGASY CHAMELEON
Chamaeleo lateralis

KNYSNA DWARF CHAMELEON
Bradypodion damaranum

367

Iguanas and Anoles

GREEN BASILISK
Basiliscus plumifrons

SHORT-HORNED LIZARD
Phrynosoma douglassii

DESERT SPINY LIZARD
Sceloporus magister

FIJIAN CRESTED IGUANA
Brachylophus vitiensis

COLLARED LIZARD
Crotaphytus collaris

COMMON BASILISK
Basiliscus basiliscus

MARINE IGUANAS

Marine iguanas of the Galápagos Islands are the only species of lizard that depend on the sea to survive.

Male marine iguanas change color to green and red to show their readiness to mate.

The cold ocean waters mean that the iguanas spend hours basking in the sun to regain enough warmth to feed again.

Geckos

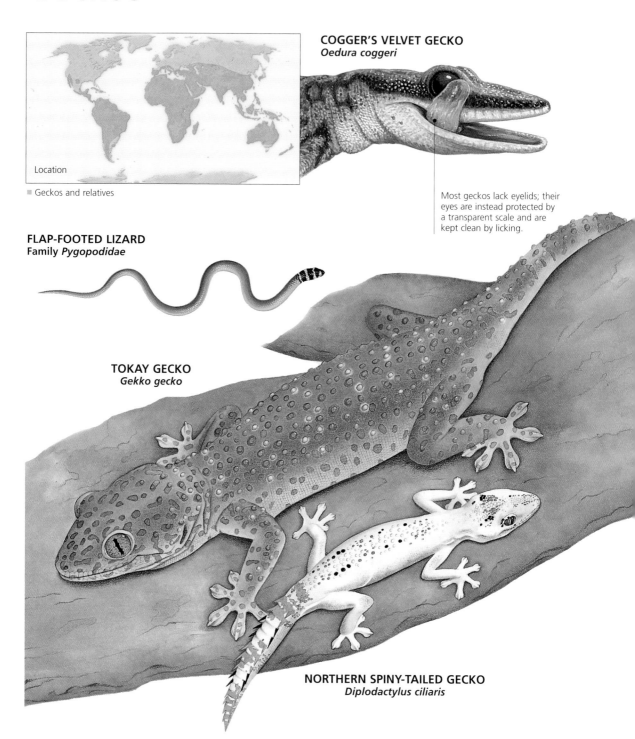

Location

■ Geckos and relatives

COGGER'S VELVET GECKO
Oedura coggeri

Most geckos lack eyelids; their eyes are instead protected by a transparent scale and are kept clean by licking.

FLAP-FOOTED LIZARD
Family *Pygopodidae*

TOKAY GECKO
Gekko gecko

NORTHERN SPINY-TAILED GECKO
Diplodactylus ciliaris

YELLOW-HEADED GECKO
Gonatodes albogularis fuscus

LEOPARD GECKO
Eublepharus macularius

GRAY'S BOW-FINGERED GECKO
Cyrtodactylus pulchellus

COMMON WONDER GECKO
Teratoscincus scincus

Geckos

FLYING GECKO
Ptychozoon **sp.**

Toes outstretched,
with webbing
between them

Flaps of skin extending along
the body enable this gecko to
glide away from predators.

BLUE-TAILED DAY GECKO
Phelsuma cepediana

RING-TAILED GECKO
Cyrtodactylus louisiadensis

SOUTHERN SPOTTED VELVET GECKO
Oedura tryoni

MOURNING GECKO
Lepidodactylus lugubris

Female mourning geckos are able to reproduce without fertilizing their eggs, and therefore without needing a male.

COMMON WALL GECKO
Tarentola mauritanica

The common wall gecko, or Moorish gecko, is one of only four species still found in Europe.

ISRAELI FAN-FINGERED GECKO
Ptyodactylus puiseuxi

Members of this rock-dwelling genus are found in Africa and the Middle East.

Beaded Lizards and Monitors

Location

- Beaded lizards
- Monitors

SENSE OF SMELL

Nasal passage

Nostril

Jacobson's organ identifies scents from tongue.

Glottis (windpipe)

Forked tongue flicks in and out of mouth to pick up chemical scents.

EMERALD TREE MONITOR
Varanus prasinus

KOMODO DRAGON
Varanus komodoensis

GOANNA NEST
Some goanna species lay their eggs in termite mounds, keeping them at a constant temperature and safe from predators.

Lizards

REPTILES

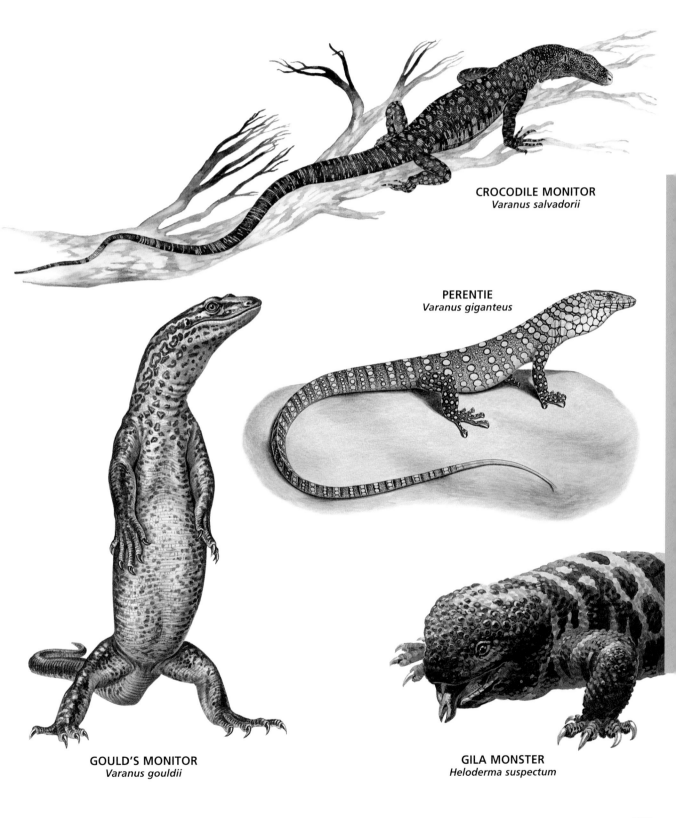

CROCODILE MONITOR
Varanus salvadorii

PERENTIE
Varanus giganteus

GOULD'S MONITOR
Varanus gouldii

GILA MONSTER
Heloderma suspectum

Skinklike Lizards

Location

■ Skinklike lizards

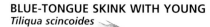

BLUE-TONGUE SKINK WITH YOUNG
Tiliqua scincoides

VIVIPAROUS LIZARD
Lacerta vivipara

SCHELTOPUSIK
Ophisaurus apodus

CHECKERED WHIPTAIL LIZARD
Cnemidophorus tesselatus

GRANITE NIGHT LIZARD
Xantusia henshawi

PINK-TONGUED SKINK
Cyclodomorphus gerrardii

SOUTH AMERICAN SKINKLIKE LIZARD
Kentropix calcarata

RAINBOW LIZARD
Cnemidophorus lemniscatus

377

Skinklike Lizards

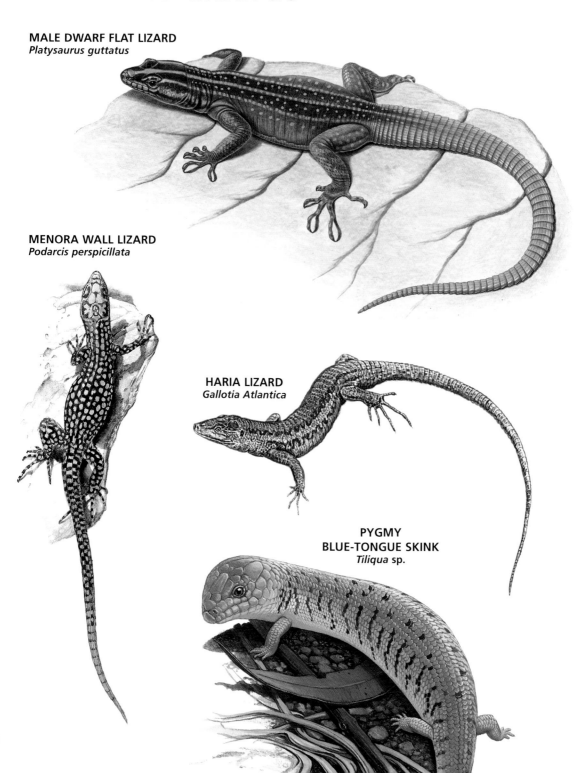

MALE DWARF FLAT LIZARD
Platysaurus guttatus

MENORA WALL LIZARD
Podarcis perspicillata

HARIA LIZARD
Gallotia Atlantica

**PYGMY
BLUE-TONGUE SKINK**
Tiliqua sp.

JUNGLE RUNNER
Ameiva ameiva

JUVENILE FIVE-LINED SKINK
Eumeces fasciatus

ITALIAN WALL LIZARD
Podarcis sicula

Snakes

EVOLUTION OF SNAKES

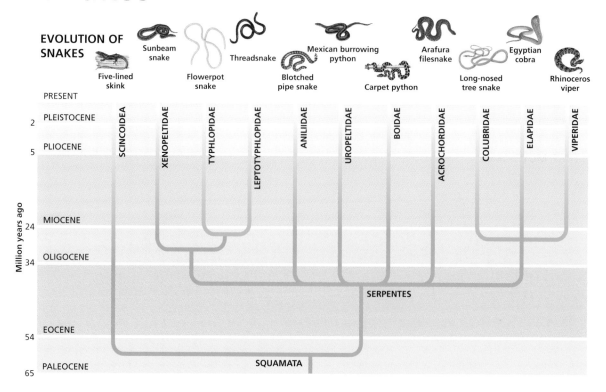

Sunbeam snake

Five-lined skink

Flowerpot snake

Threadsnake

Mexican burrowing python

Blotched pipe snake

Carpet python

Arafura filesnake

Long-nosed tree snake

Egyptian cobra

Rhinoceros viper

Million years ago

	PRESENT
2	PLEISTOCENE
5	PLIOCENE
24	MIOCENE
34	OLIGOCENE
54	EOCENE
65	PALEOCENE

SCINCOIDEA
XENOPELTIDAE
TYPHLOPIDAE
LEPTOTYPHLOPIDAE
ANILIIDAE
UROPELTIDAE
BOIDAE
ACROCHORDIDAE
COLUBRIDAE
ELAPIDAE
VIPERIDAE

SERPENTES

SQUAMATA

SNAKE ANATOMY

Hinged fang can be folded back into the mouth.

Venom duct passes venom through to the hollow fang.

Single, narrow lung runs nearly the length of the body.

Liver

Forked tongue senses smells in the air.

Venom gland

Snakes can have between 150 and 450 pairs of ribs.

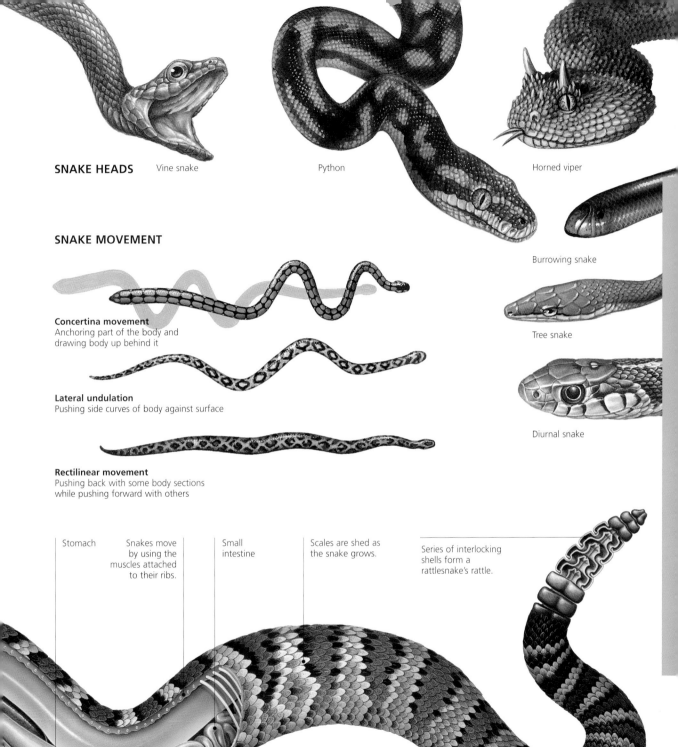

SNAKE HEADS
Vine snake
Python
Horned viper

Burrowing snake

Tree snake

Diurnal snake

SNAKE MOVEMENT

Concertina movement
Anchoring part of the body and
drawing body up behind it

Lateral undulation
Pushing side curves of body against surface

Rectilinear movement
Pushing back with some body sections
while pushing forward with others

Stomach

Snakes move
by using the
muscles attached
to their ribs.

Small
intestine

Scales are shed as
the snake grows.

Series of interlocking
shells form a
rattlesnake's rattle.

Wormsnakes and Pipe Snakes

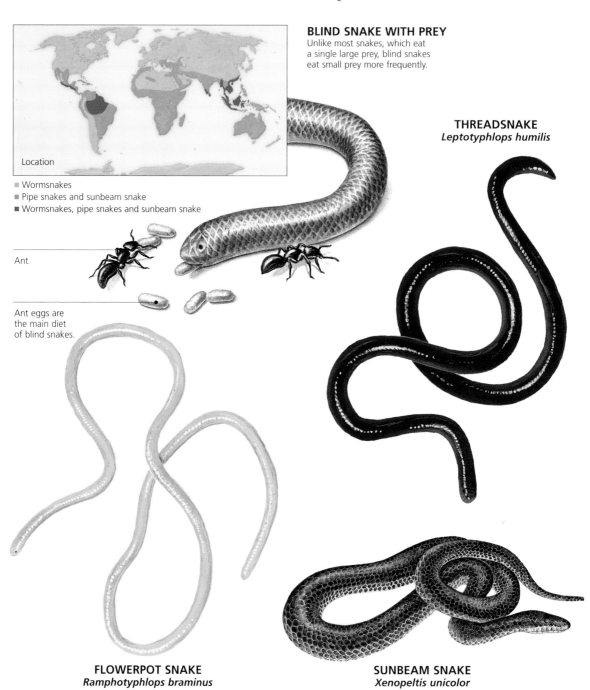

Location

- Wormsnakes
- Pipe snakes and sunbeam snake
- Wormsnakes, pipe snakes and sunbeam snake

BLIND SNAKE WITH PREY
Unlike most snakes, which eat a single large prey, blind snakes eat small prey more frequently.

Ant

Ant eggs are the main diet of blind snakes.

THREADSNAKE
Leptotyphlops humilis

FLOWERPOT SNAKE
Ramphotyphlops braminus

SUNBEAM SNAKE
Xenopeltis unicolor

Snakes

REPTILES

MEXICAN BURROWING PYTHON
Loxocemus bicolor

**SOUTH AMERICAN CORAL
PIPE SNAKE**
Anilius scytale

**DRUMMOND HAY'S
EARTH SNAKE**
Rhinophis drummondhayi

BLOTCHED PIPE SNAKE
Cylindrophus maculatus

RED CYLINDER SNAKE
Cylindrophis rufus

Boas and Pythons

Location
- Boas
- Pythons
- Boas and pythons

ELASTIC JAWS

Resting jaw

Extended jaw
Elastic hinges in the boa's jaw enable the mouth to stretch wide enough to consume prey twice the size of the boa's head.

PYTHON DIGESTING MEAL
Resting in the heat of the sun allows a python to digest its meal more rapidly.

PYTHON ENGULFING WILD PIG
Digesting prey this size may take weeks or months.

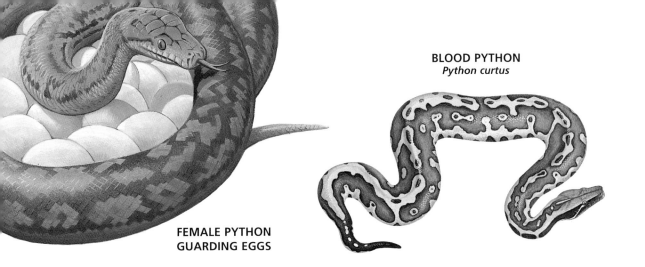

**FEMALE PYTHON
GUARDING EGGS**

BLOOD PYTHON
Python curtus

SIZE COMPARISON

Anaconda (species of boa): 33 feet (10 m)

Boa constrictor: 14½ feet (4.5 m)

Eastern diamondback rattlesnake: 7 feet (2.2 m)

Yellow-bellied sea snake: 2½ feet (0.8 m)

BRAZILIAN RAINBOW BOA
Epicrates cenchria cenchria

CARPET PYTHON
Morelia spilota

Colubrid Snakes

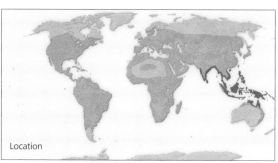

Location

- Colubrid snakes
- Colubrid snakes and filesnakes

WHITE-BELLIED MANGROVE SNAKE
Fordonia leucobalia

ARAFURA FILESNAKE
Acrochordus arafurae

LONG-NOSED TREE SNAKE
Ahaetulla nasuta

AFRICAN TWIG SNAKE
Thelotornis capensis

LITTLE FILESNAKE
Acrochordus granulatus

COMMON GRASS SNAKE
Natrix natrix

SPOTTED HOUSE SNAKE
Lamprophis guttatus

COMMON KING SNAKE
Lampropeltis getulus

TOAD-EATER SNAKE
Xenodon rabdocephalus

Colubrid Snakes

**BLUNT-HEADED
TREE SNAKE**
Imantodes cenchoa

**SPOTTED HARLEQUIN
SNAKE**
Homoroselaps lacteus

MANDARIN RATSNAKE
Elaphe mandarina

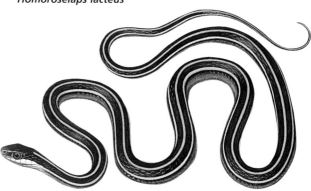

EASTERN RIBBON SNAKE
Thamnophis sauritus sauritus

GREEN VINE SNAKE
Oxybelis sp.

Even though green vine snakes can grow to 7 feet (2 m) in length, their bodies are no more than ½ inch (1.3 cm) wide.

Their slender bodies can curl easily around branches and vines as they search for small birds in nests.

Elapid Snakes and Vipers

TYPES OF FANGS

Colubrid snakes
Rear, fixed, grooved fangs

Elapid snakes
Front, fixed, hollow fangs

Vipers
Front, swinging, hollow fangs

RATTLESNAKE HEAD
Like vipers, rattlesnakes have swinging fangs, and can sense heat so accurately they can strike prey in total darkness.

SPITTING COBRA

Lacking the lips to 'spit' in the traditional sense, spitting cobras force venom through openings in their fangs onto would-be attackers.

The cobra's hood flap is made from skin only, which extends as the elongated ribs expand when it feels angry or threatened.

COBRA SKELETON

Elongated ribs

SIDEWINDER LOCOMOTION

The sidewinder moves across hot, loose sand by anchoring its head and tail, lifting its trunk sideways across the ground. The head and tail then join the trunk.

Tracks left in the sand dune

RATTLESNAKE RATTLE

Body scales

Interlocking shells

The rattle is used to distract prey, mesmerizing would-be attackers until the snake is ready to strike.

Elapid Snakes

Location

- Elapid snakes
- Coral snakes and sea snakes

BLUE CORAL SNAKE HEAD
Maticora bivirgata

The venom duct of the blue coral snake extends almost a third of the way along the snake's body.

EGYPTIAN COBRA
Naja haje

YELLOW-LIPPED SEA KRAIT
Laticauda colubrina

ARIZONA CORAL SNAKE
*Micruroides euryxanthus
euryxanthus*

VENOMOUS
IMITATORS

Mayan coral snake
Venomous

False coral snake
Non-venomous

The yellow banding
on tail is the only
visible difference.

TURTLE-HEADED SEA SNAKE
Emydocephalus annulatus

**YELLOW-BELLIED
SEA SNAKE**
Pelamis platurus

RED-BELLIED BLACK SNAKE
Pseudechis porphyriacus

BLACK MAMBA
Dendroaspis polylepis

Vipers

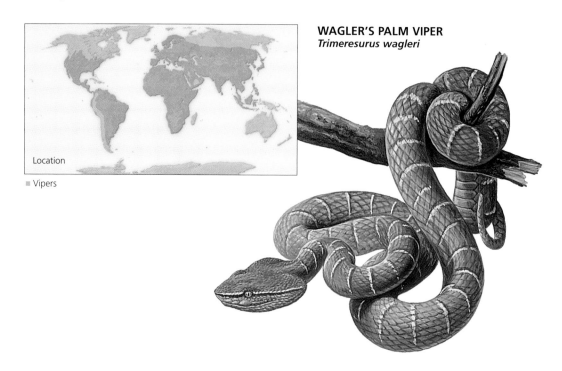

Location

■ Vipers

WAGLER'S PALM VIPER
Trimeresurus wagleri

URUTU
Bothrops alternatus

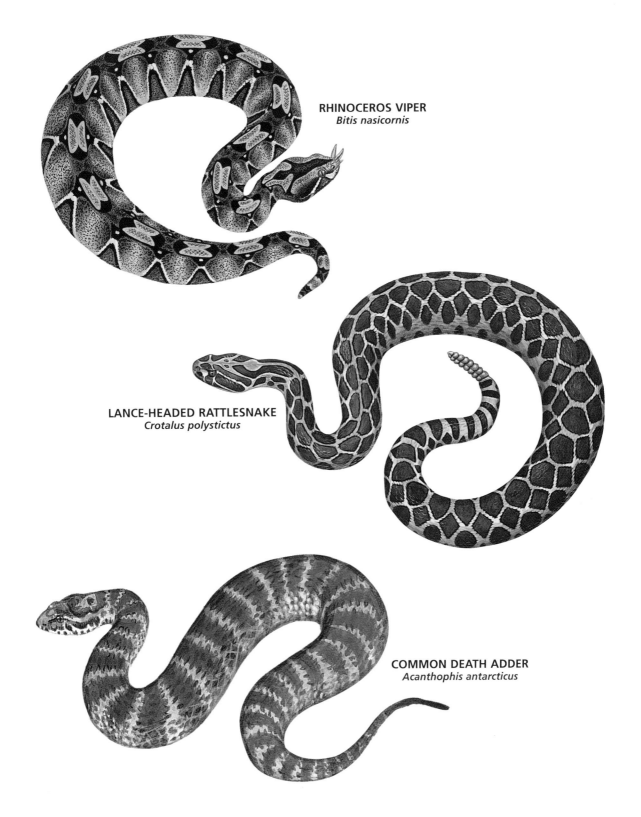

RHINOCEROS VIPER
Bitis nasicornis

LANCE-HEADED RATTLESNAKE
Crotalus polystictus

COMMON DEATH ADDER
Acanthophis antarcticus

INTRODUCING AMPHIBIANS

SALAMANDERS AND NEWTS

CAECILIANS

FROGS AND TOADS

Classifying Amphibians

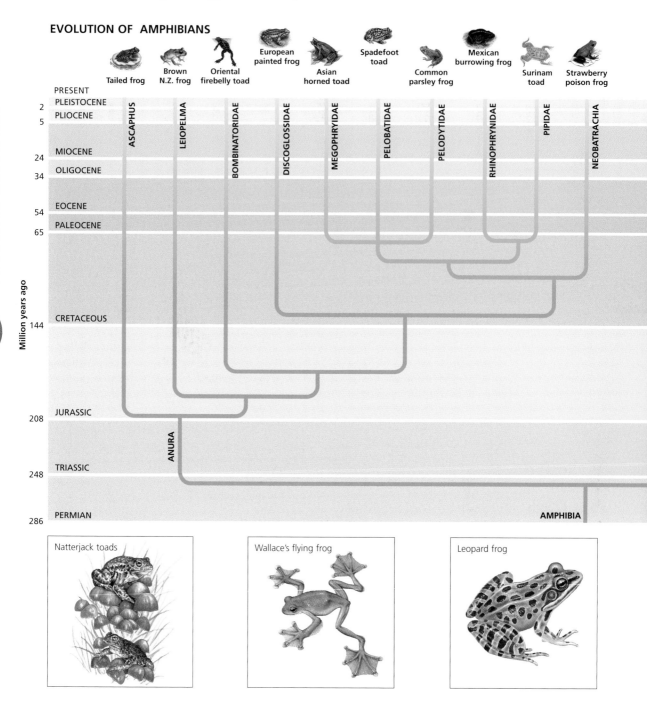

EVOLUTION OF AMPHIBIANS

Tailed frog · Brown N.Z. frog · Oriental firebelly toad · European painted frog · Asian horned toad · Spadefoot toad · Common parsley frog · Mexican burrowing frog · Surinam toad · Strawberry poison frog

ASCAPHUS · LEIOPELMA · BOMBINATORIDAE · DISCOGLOSSIDAE · MEGOPHRYIDAE · PELOBATIDAE · PELODYTIDAE · RHINOPHRYNIDAE · PIPIDAE · NEOBATRACHIA

ANURA

AMPHIBIA

Million years ago

PRESENT
PLEISTOCENE — 2
PLIOCENE — 5
MIOCENE — 24
OLIGOCENE — 34
EOCENE — 54
PALEOCENE — 65
CRETACEOUS — 144
JURASSIC — 208
TRIASSIC — 248
PERMIAN — 286

Introducing Amphibians

AMPHIBIANS

Natterjack toads

Wallace's flying frog

Leopard frog

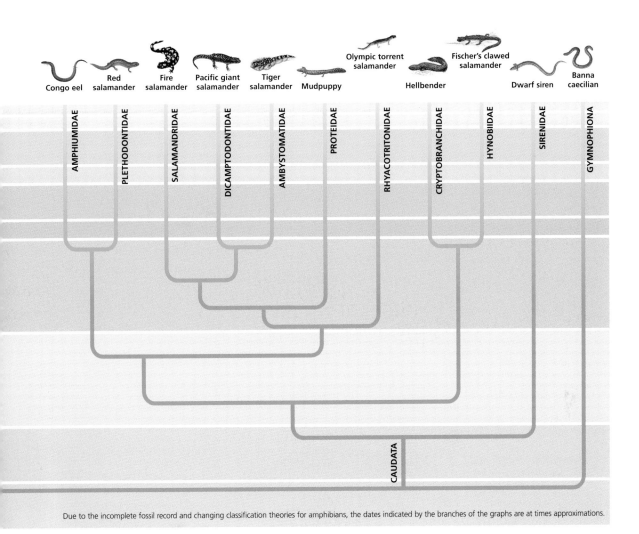

Congo eel Red salamander Fire salamander Pacific giant salamander Tiger salamander Mudpuppy Olympic torrent salamander Hellbender Fischer's clawed salamander Dwarf siren Banna caecilian

AMPHIUMIDAE
PLETHODONTIDAE
SALAMANDRIDAE
DICAMPTODONTIDAE
AMBYSTOMATIDAE
PROTEIDAE
RHYACOTRITONIDAE
CRYPTOBRANCHIDAE
HYNOBIIDAE
SIRENIDAE
GYMNOPHIONA

CAUDATA

Due to the incomplete fossil record and changing classification theories for amphibians, the dates indicated by the branches of the graphs are at times approximations.

Poison frog in bromeliad

Mexican axolotyl

Terrestrial caecilian

Salamanders and Newts

EVOLUTION OF SALAMANDERS AND NEWTS

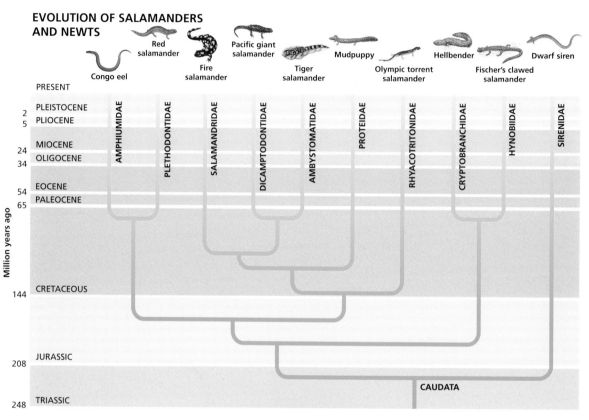

SALAMANDER LIFECYCLE

While lifecycles vary greatly across salamander species, the basic lifecycle, with its aquatic and terrestrial stages, is shown.

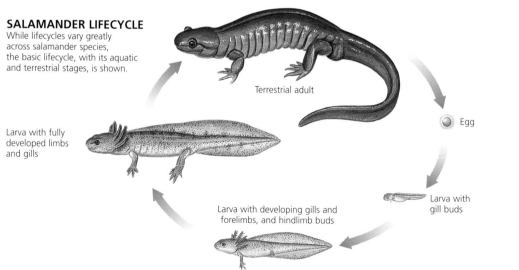

Terrestrial adult

Egg

Larva with fully developed limbs and gills

Larva with gill buds

Larva with developing gills and forelimbs, and hindlimb buds

FIRE SALAMANDER

FIRE SALAMANDER SKELETON

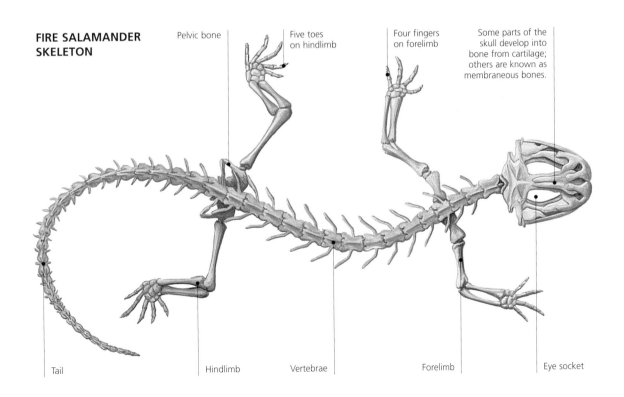

Pelvic bone

Five toes on hindlimb

Four fingers on forelimb

Some parts of the skull develop into bone from cartilage; others are known as membraneous bones.

Tail

Hindlimb

Vertebrae

Forelimb

Eye socket

Salamanders and Newts

Location

■ Salamanders and newts

FIRE SALAMANDER
Salamandra salamandra

FISCHER'S CLAWED SALAMANDER
Onychodactylus fischeri

BLUE-SPOTTED SALAMANDER
Ambystoma laterale

PACIFIC GIANT SALAMANDER
Dicamptodon ensatus

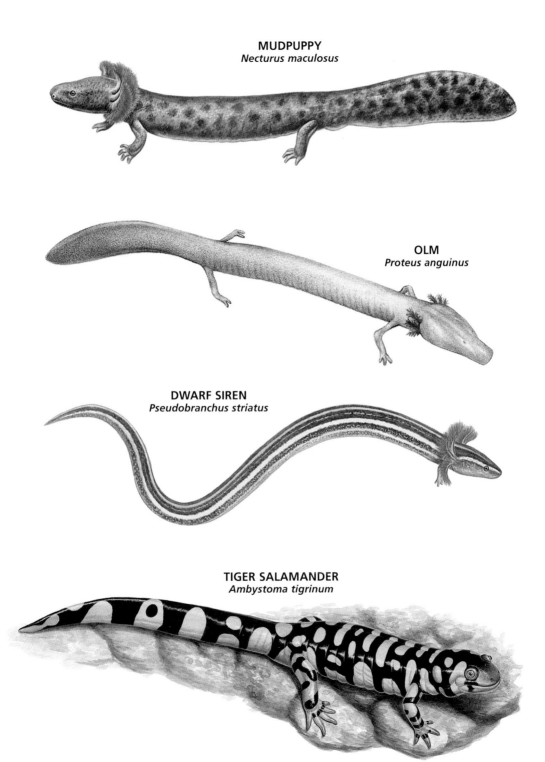

MUDPUPPY
Necturus maculosus

OLM
Proteus anguinus

DWARF SIREN
Pseudobranchus striatus

TIGER SALAMANDER
Ambystoma tigrinum

Salamanders and Newts

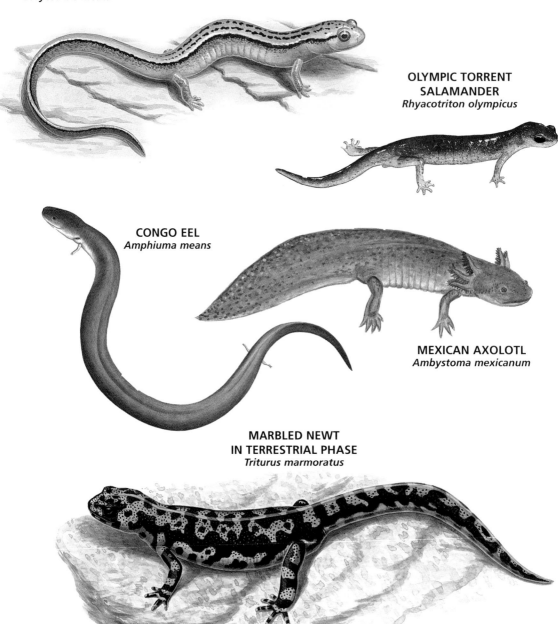

TWO-LINED SALAMANDER
Eurycea bislineata

**OLYMPIC TORRENT
SALAMANDER**
Rhyacotriton olympicus

CONGO EEL
Amphiuma means

MEXICAN AXOLOTL
Ambystoma mexicanum

**MARBLED NEWT
IN TERRESTRIAL PHASE**
Triturus marmoratus

JAPANESE GIANT SALAMANDER
Andrias japonicus

**ALPINE NEWT
IN AQUATIC PHASE**
Triturus alpestris

RED SALAMANDER
Pseudotriton ruber

HELLBENDER
Cryptobranchus alleganiensis

Caecilians

Location

■ Caecilians

RINGED CAECILIAN
Siphonops annulatus

BANNA CAECILIAN
Ichthyophis bannanicus

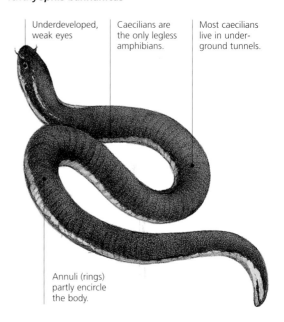

Underdeveloped, weak eyes

Caecilians are the only legless amphibians.

Most caecilians live in underground tunnels.

Annuli (rings) partly encircle the body.

ADAPTATIONS FOR BURROWING

Family Rhinatrematidae
Epicrionops petersi

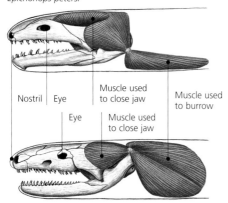

Nostril

Eye

Muscle used to close jaw

Muscle used to burrow

Eye

Muscle used to close jaw

Family Ichthyophiidae
Ichthyophis glutinosus

Family Caecilidae
Microcaecilia rabei

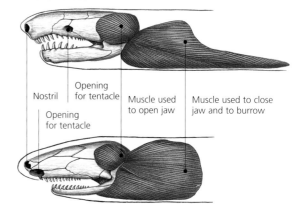

Nostril

Opening for tentacle

Muscle used to open jaw

Muscle used to close jaw and to burrow

Opening for tentacle

Family Scolecomorphidae
Crotaphatrema lamottei

AQUATIC CAECILIAN
Typhlonectes natans

SOUTHEAST ASIAN CAECILIAN
Ichthyophis kohtaoensis

TERRESTRIAL CAECILIAN
Dermophis mexicanus

SAO TOMÉ CAECILIAN
Schistometopum thomense

CAYENNE CAECILIAN
Typhlonectes compressicauda

Frogs and Toads

EVOLUTION OF FROGS AND TOADS

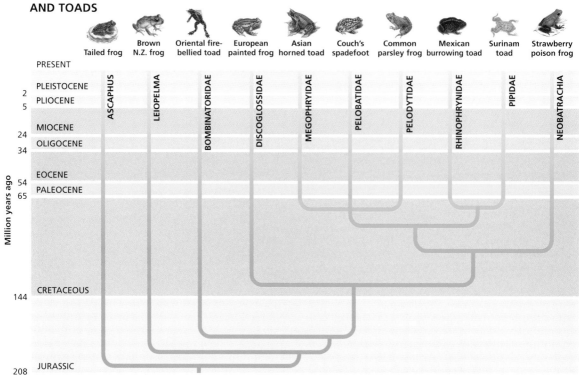

Tailed frog | Brown N.Z. frog | Oriental fire-bellied toad | European painted frog | Asian horned toad | Couch's spadefoot | Common parsley frog | Mexican burrowing toad | Surinam toad | Strawberry poison frog

ASCAPHUS | LEIOPELMA | BOMBINATORIDAE | DISCOGLOSSIDAE | MEGOPHRYIDAE | PELOBATIDAE | PELODYTIDAE | RHINOPHRYNIDAE | PIPIDAE | NEOBATRACHIA

Million years ago

PRESENT
PLEISTOCENE
PLIOCENE — 2, 5
MIOCENE — 24
OLIGOCENE — 34
EOCENE — 54
PALEOCENE — 65
CRETACEOUS — 144
JURASSIC — 208

TREEFROG

TREEFROG SKELETON

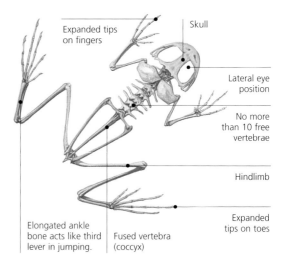

Expanded tips on fingers

Skull

Lateral eye position

No more than 10 free vertebrae

Hindlimb

Expanded tips on toes

Elongated ankle bone acts like third lever in jumping.

Fused vertebra (coccyx)

410

FROG LIFECYCLE

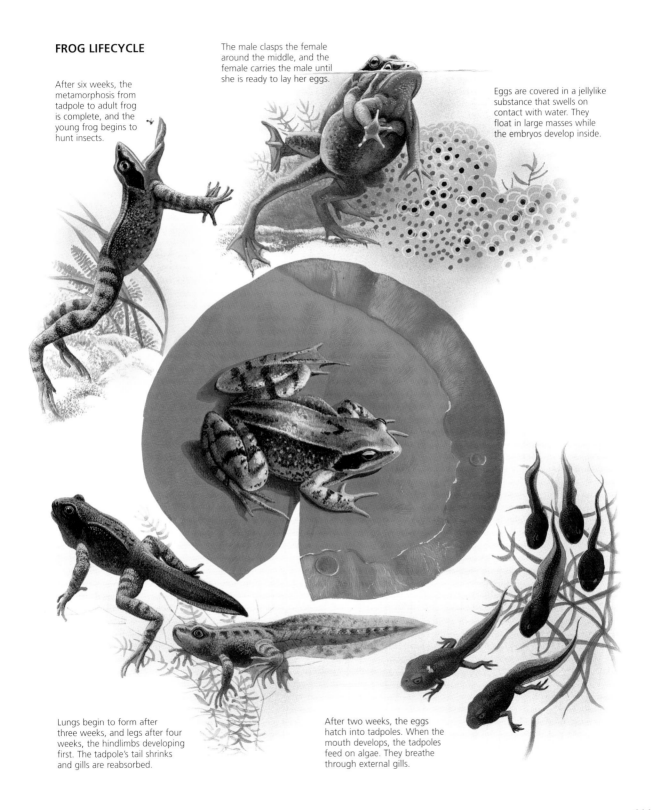

The male clasps the female around the middle, and the female carries the male until she is ready to lay her eggs.

After six weeks, the metamorphosis from tadpole to adult frog is complete, and the young frog begins to hunt insects.

Eggs are covered in a jellylike substance that swells on contact with water. They float in large masses while the embryos develop inside.

Lungs begin to form after three weeks, and legs after four weeks, the hindlimbs developing first. The tadpole's tail shrinks and gills are reabsorbed.

After two weeks, the eggs hatch into tadpoles. When the mouth develops, the tadpoles feed on algae. They breathe through external gills.

Primitive Frogs

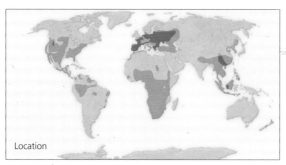

Location

■ Primitive frogs
■ Pipids and transitional frogs
■ Primitive frogs, pipids and transitional frogs

YELLOW-BELLIED TOAD
Bombina variegata

SURINAM TOAD
Pipa pipa

ASIAN HORNED TOAD
Megophrys nasuta

TAILED FROG
Ascaphus truei

MEXICAN BURROWING TOAD
Rhinophrynus dorsalis

ORIENTAL FIRE-BELLIED TOAD
Bombina orientalis

CAPE CLAWED FROG
Xenopus gilli

NEW ZEALAND FROG
Family Leiopelmatidae

BROWN NEW ZEALAND FROG
Leiopelma hamiltoni

COMMON PARSLEY FROG
Pelodytes punctatus

EUROPEAN PAINTED FROG
Discoglossus pictus

COUCH'S SPADEFOOT
Scaphiopus couchii

413

American and Australasian Frogs

Location

- Leptodactylid frogs (American)
- Myobatrachid frogs (Australasian)

SOUTHERN PLATYPUS FROG
Rheobatrachus silus

GIANT BULLFROG
Limnodynastes interioris

CRUCIFIX TOAD
Notaden bennettii

WESTERN BARKING FROG
Eleutherodactylus augusti

WESTERN MARSH FROG
Heleioporus barycragus

CORROBOREE FROG
Pseudophryne corroboree

ORNATE HORNED TOAD
Ceratophrys ornata

SPOTTED GRASS FROG
Limnodynastes tasmaniensis

SOUTH AMERICAN BULLFROG
Leptodactylus pentadactylus

SCHMIDT'S FOREST FROG
Hydrolaetare schmidti

415

Toads, Treefrogs and Glass Frogs

Location

- Toads and harlequin frogs
- Treefrogs and glass frogs
- Toads, harlequin frogs, treefrogs and glass frogs

**EVERETT'S ASIAN
TREE TOAD**
Pedostibes everetti

VARIABLE HARLEQUIN FROG
Atelopus varius

LEOPARD TOAD
Bufo pardalis

**GREEN AND GOLDEN
BELL FROG**
Litoria aurea

416

ASIATIC CLIMBING TOAD
Pedostibes hosii

DAINTY GREEN TREEFROG
Litoria gracilenta

GLASS FROG
Hylinobatrachium fleischmanni

BURROWING TREEFROG
Pternohyla fodiens

EUROPEAN GREEN TOAD
Bufo viridis

RED-EYED TREEFROG
Agalychnis callidryas

True Frogs and Poison Frogs

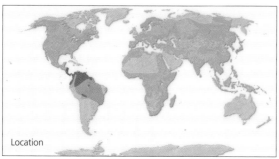

Location

- True frogs
- Poison frogs
- True frogs and poison frogs

ORNATE BURROWING FROG
Hildebrandtia ornata

FUNEREAL POISON FROG
Phyllobates lugubris

SOLOMON ISLANDS TREEFROG
Platymantis guppyi

NORTHERN LEOPARD FROG
Rana sp.

ORANGE AND BLACK POISON FROG
Dendrobates leucomelas

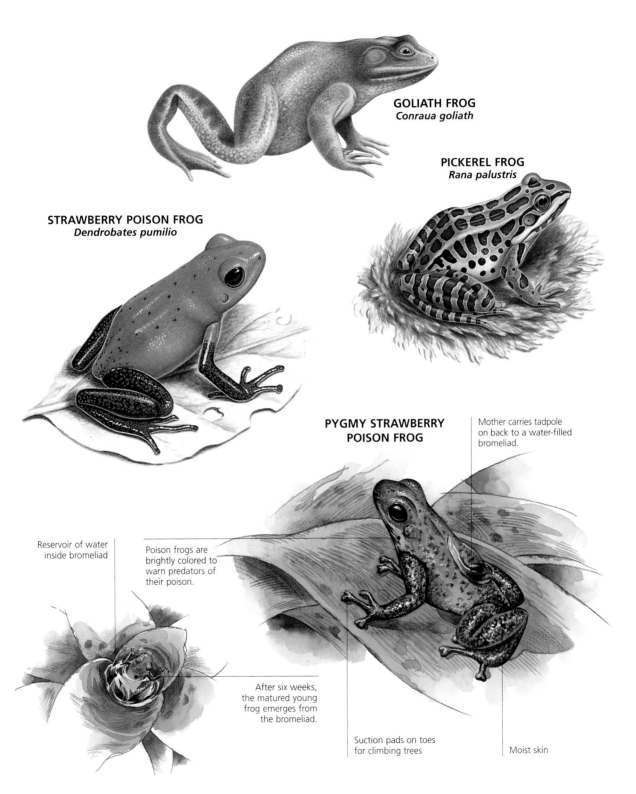

GOLIATH FROG
Conraua goliath

PICKEREL FROG
Rana palustris

STRAWBERRY POISON FROG
Dendrobates pumilio

PYGMY STRAWBERRY POISON FROG

Mother carries tadpole on back to a water-filled bromeliad.

Reservoir of water inside bromeliad

Poison frogs are brightly colored to warn predators of their poison.

After six weeks, the matured young frog emerges from the bromeliad.

Suction pads on toes for climbing trees

Moist skin

Other Frogs and Toads

Location

■ Other frogs and toads

GOLD FROG
Brachycephalus ephippium

CAPE GHOST FROG
Heleophryne purcelli

SEYCHELLES FROG
Sooglossus sechellensis

ASIAN PAINTED FROG
Kaloula pulchra

SENEGAL RUNNING FROG
Kassina senegalensis

DARWIN'S FROG
Rhinoderma darwinii

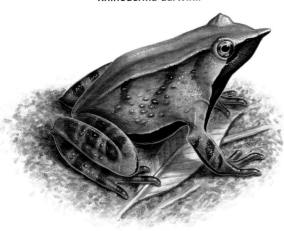

WALLACE'S FLYING FROG
Rhacophorus nigropalmatus

EASTERN
NARROW-MOUTHED TOAD
Gastrophryne carolinensis

PARADOX FROG
Pseudis paradoxa

PAINTED REED FROG
Hyperolius marmoratus

RED-BANDED CREVICE
CREEPER
Phrynomerus bifasciatus

Classifying Fishes

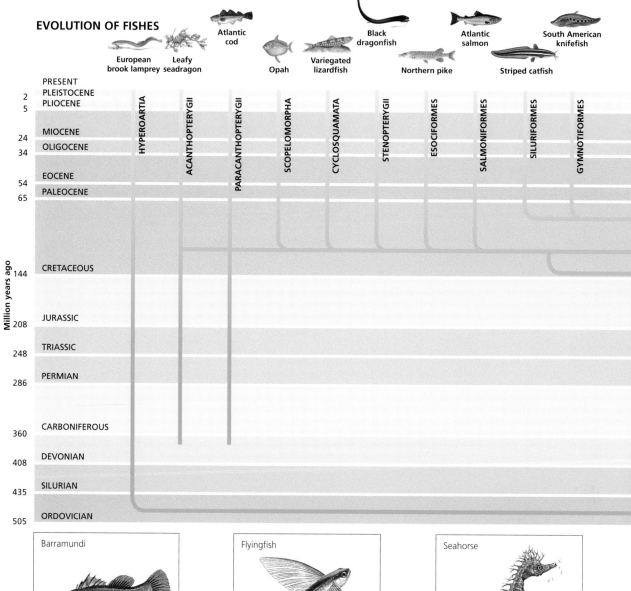

EVOLUTION OF FISHES

European brook lamprey

Leafy seadragon

Atlantic cod

Opah

Variegated lizardfish

Black dragonfish

Northern pike

Atlantic salmon

Striped catfish

South American knifefish

(vertical axis labels:) Introducing Fishes — FISHES — Million years ago

Period / Epoch	Million years ago
PRESENT	
PLEISTOCENE	2
PLIOCENE	5
MIOCENE	24
OLIGOCENE	34
EOCENE	54
PALEOCENE	65
CRETACEOUS	144
JURASSIC	208
TRIASSIC	248
PERMIAN	286
CARBONIFEROUS	360
DEVONIAN	408
SILURIAN	435
ORDOVICIAN	505

(column labels:) HYPEROARTIA · ACANTHOPTERYGII · PARACANTHOPTERYGII · SCOPELOMORPHA · CYCLOSQUAMATA · STENOPTERYGII · ESOCIFORMES · SALMONIFORMES · SILURIFORMES · GYMNOTIFORMES

Barramundi

Flyingfish

Seahorse

426

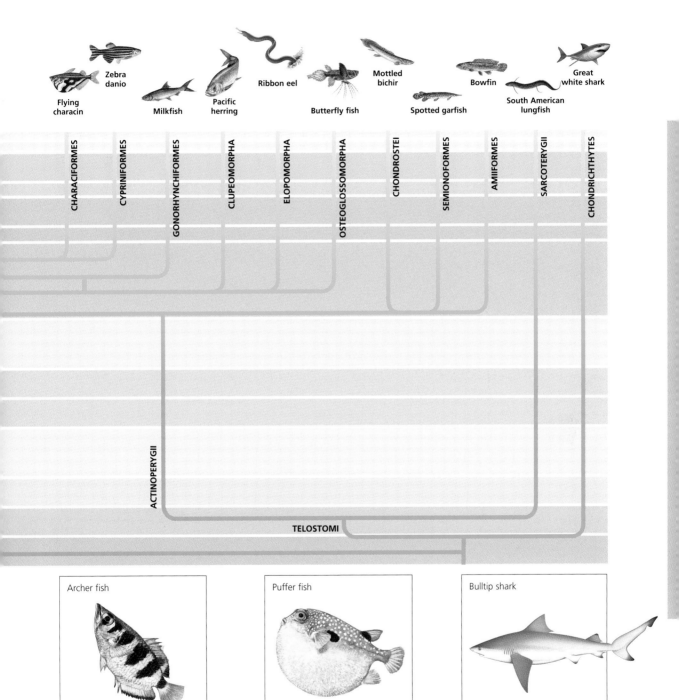

Flying characin

Zebra danio

Milkfish

Pacific herring

Ribbon eel

Butterfly fish

Mottled bichir

Spotted garfish

Bowfin

South American lungfish

Great white shark

CHARACIFORMES

CYPRINIFORMES

GONORHYNCHIFORMES

CLUPEOMORPHA

ELOPOMORPHA

OSTEOGLOSSOMORPHA

CHONDROSTEI

SEMIONOFORMES

AMIIFORMES

SARCOTERYGII

CHONDRICHTHYTES

ACTINOPERYGII

TELOSTOMI

Archer fish

Puffer fish

Bulltip shark

Fish Characteristics

BONY FISH

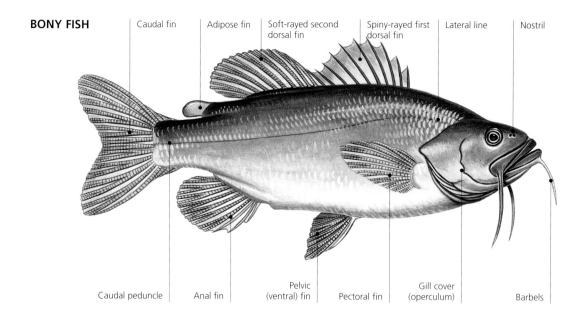

Caudal fin | Adipose fin | Soft-rayed second dorsal fin | Spiny-rayed first dorsal fin | Lateral line | Nostril

Caudal peduncle | Anal fin | Pelvic (ventral) fin | Pectoral fin | Gill cover (operculum) | Barbels

CARTILAGINOUS FISH

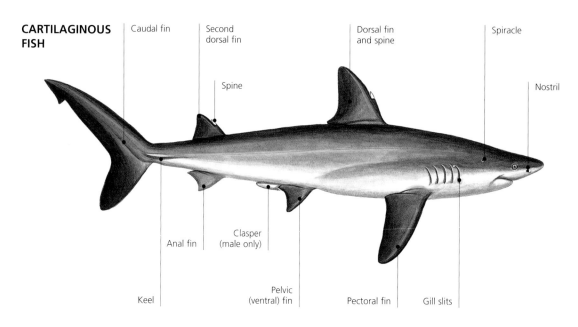

Caudal fin | Second dorsal fin | Dorsal fin and spine | Spiracle

Spine | Nostril

Anal fin | Clasper (male only)

Keel | Pelvic (ventral) fin | Pectoral fin | Gill slits

FISH SCALES

Ctenoid: bony fishes

Cycloid: bony fishes

Ganoid: armored fishes

Placoid: sharks

FISH RESPIRATION

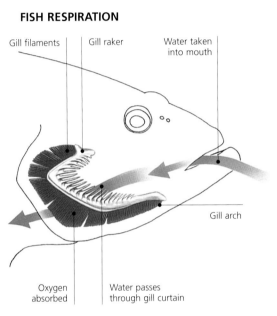

Gill filaments

Gill raker

Water taken into mouth

Gill arch

Oxygen absorbed

Water passes through gill curtain

SHARK RESPIRATION

Gill arch

Gill opening

Ventral aorta

Gill filaments

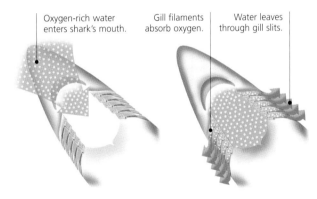

Oxygen-rich water enters shark's mouth.

Gill filaments absorb oxygen.

Water leaves through gill slits.

FISH TAILS

Butterflyfish

Anemone fish

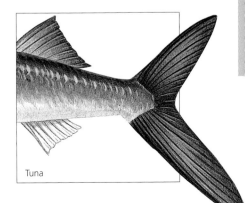

Tuna

Evolution of Fishes

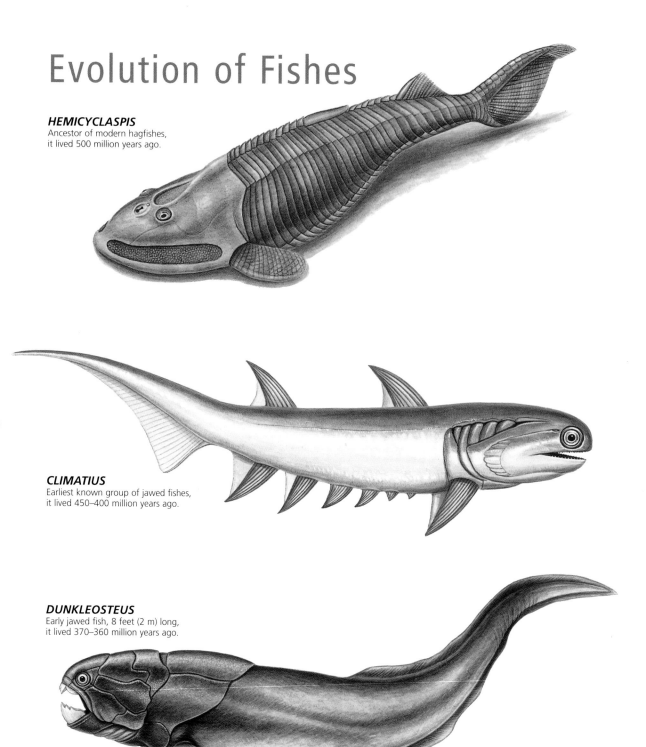

HEMICYCLASPIS
Ancestor of modern hagfishes,
it lived 500 million years ago.

CLIMATIUS
Earliest known group of jawed fishes,
it lived 450–400 million years ago.

DUNKLEOSTEUS
Early jawed fish, 8 feet (2 m) long,
it lived 370–360 million years ago.

CLADOSELACHE
Ancestor of modern sharks,
it lived 380 million years ago.

CHEIROLEPIS
Possibly an ancestor of sturgeons and
bichirs, it lived 380 million years ago.

EUSTHENOPTERON
Lobe-finned fish with simple lungs,
it lived 360 million years ago.

Jawless Fishes

Location

■ Lampreys and hagfishes

SEA LAMPREY
Petromyzon marinus

EUROPEAN BROOK LAMPREY
Lampetra planeri

EUROPEAN RIVER LAMPREY
Lampetra fluviatilis

ATLANTIC HAGFISH
Myxine glutinosa

PACIFIC HAGFISH
Eptatretus stouti

HAGFISH KNOTS

By tying themselves in a knot, and bracing themselves against a carcass, limbless and jawless hagfishes gain enough leverage to tear off pieces of flesh.

Sharks

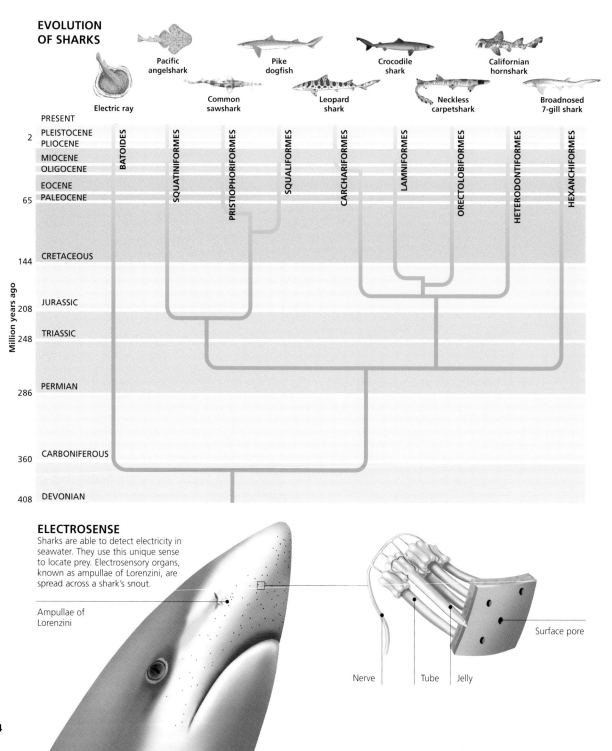

EVOLUTION OF SHARKS

Pacific angelshark

Electric ray

Pike dogfish

Common sawshark

Crocodile shark

Leopard shark

Californian hornshark

Neckless carpetshark

Broadnosed 7-gill shark

Million years ago

PRESENT

2 — PLEISTOCENE / PLIOCENE

MIOCENE / OLIGOCENE

EOCENE / PALEOCENE

65

144 — CRETACEOUS

208 — JURASSIC

248 — TRIASSIC

286 — PERMIAN

360 — CARBONIFEROUS

408 — DEVONIAN

BATOIDES

SQUATINIFORMES

PRISTIOPHORIFORMES

SQUALIFORMES

CARCHARIFORMES

LAMNIFORMES

ORECTOLOBIFORMES

HETERODONTIFORMES

HEXANCHIFORMES

ELECTROSENSE

Sharks are able to detect electricity in seawater. They use this unique sense to locate prey. Electrosensory organs, known as ampullae of Lorenzini, are spread across a shark's snout.

Ampullae of Lorenzini

Surface pore

Nerve   Tube   Jelly

SHARK REPRODUCTION

Hornsharks

Hornsharks lay oval egg cases ringed with screwlike ridges. The mother wedges each case firmly into a crevice.

Swellsharks

Swellsharks lay flat, rectangular egg cases.

A growing shark and the yolk that feeds it.

The newborn shark begins to emerge.

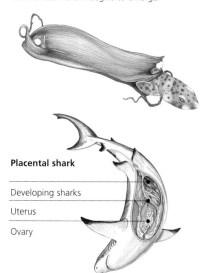

Placental shark

Developing sharks

Uterus

Ovary

TYPES OF TEETH

Blue shark

Tiger shark

Hornshark

Great white shark

Shortfin mako

ANATOMY OF FEEDING

In its resting position, the shark's jaw lies just under its brain case.

As the shark goes to grab its prey, the upper jaw slides forward. The lower jaw drops.

Muscle contractions force the jaw out of the mouth to give the shark a good grip.

Angel Sharks, Sawfishes and Dogfish Sharks

Location

- Sawfishes and dogfish sharks
- Sawfishes, dogfish sharks and angel sharks

SPINED PYGMY SHARK
Squaliolus laticaudus

COOKIE-CUTTER SHARK
Isistius brasiliensis

SMALLFIN GULPER SHARK
Centrophorus moluccensis

SPINY DOGFISH
Squalus acanthias

Sharks

FISHES

BLACKBELLY LANTERNSHARK
Etmopterus lucifer

PACIFIC ANGEL SHARK
Squatina californica

PRICKLY DOGFISH
Oxynotus bruniensis

GREENLAND SHARK
Somniosus microcephalus

BRAMBLE SHARK
Echinorhinus brucus

LONGNOSE SAWFISH
Pristiophorus cirratus

Ground Sharks

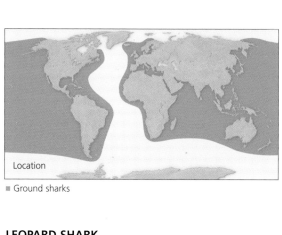

Location

■ Ground sharks

TYPES OF TAILS

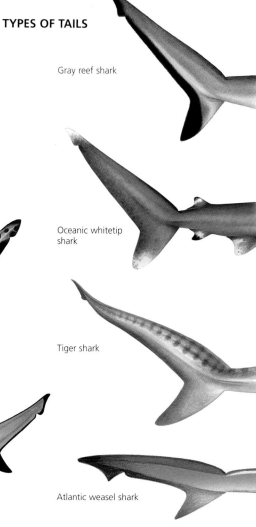

Gray reef shark

Oceanic whitetip shark

Tiger shark

Atlantic weasel shark

Swell shark

LEOPARD SHARK
Triakis semifasciata

BLACKTIP REEF SHARK
Carcharhinus melanopterus

GRACEFUL CATSHARK
Proscyllium habereri

SCALLOPED HAMMERHEAD
Sphyrna lewini

HEAD SHAPES

Winghead

Scalloped hammerhead

Bonnethead

HEAD HUNTING

Stingrays are the preferred prey of great hammerhead sharks.

The shark's eyes are located on the ends of the wings on their head.

It bites large chunks out of its victim, devouring it piece by piece.

The shark pins a stingray to the sand with the side of its head.

Mackerel Sharks

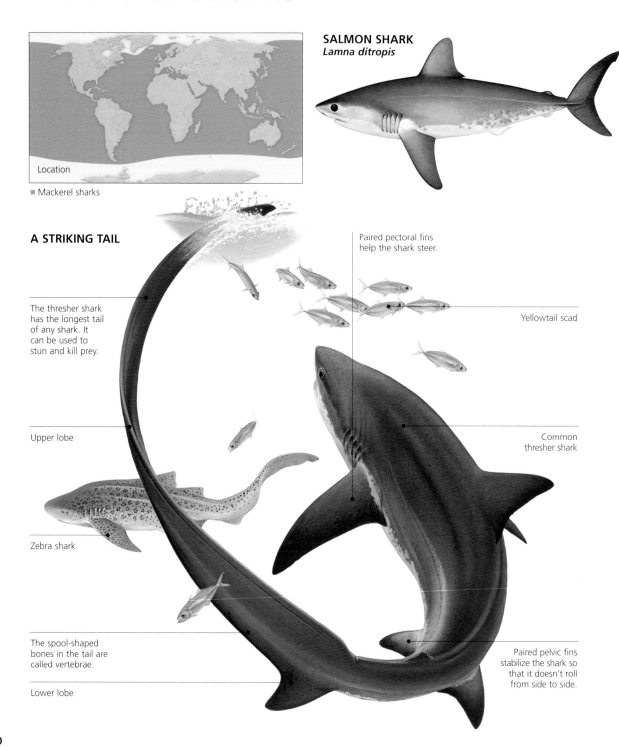

Location

■ Mackerel sharks

SALMON SHARK
Lamna ditropis

A STRIKING TAIL

Paired pectoral fins
help the shark steer.

The thresher shark
has the longest tail
of any shark. It
can be used to
stun and kill prey.

Yellowtail scad

Upper lobe

Common
thresher shark

Zebra shark

The spool-shaped
bones in the tail are
called vertebrae.

Paired pelvic fins
stabilize the shark so
that it doesn't roll
from side to side.

Lower lobe

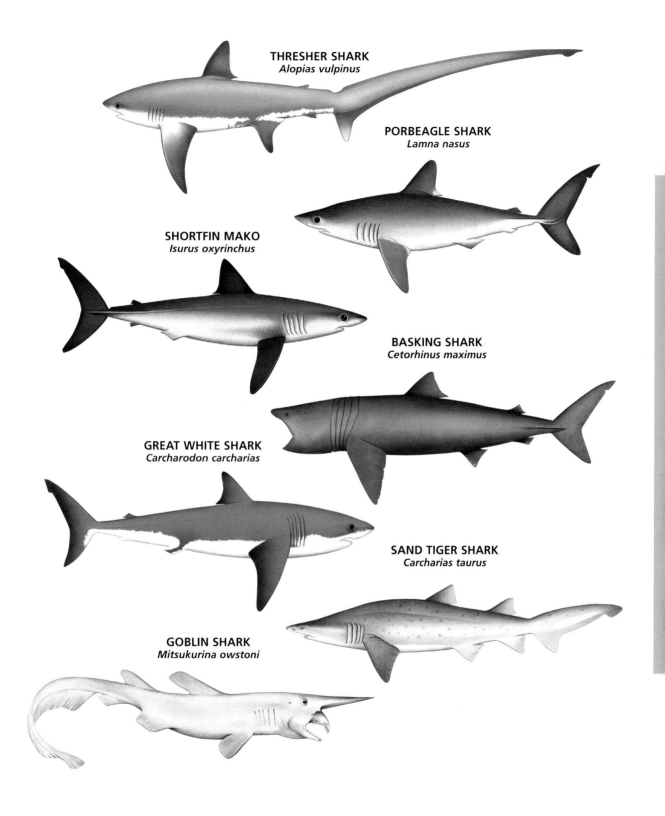

THRESHER SHARK
Alopias vulpinus

PORBEAGLE SHARK
Lamna nasus

SHORTFIN MAKO
Isurus oxyrinchus

BASKING SHARK
Cetorhinus maximus

GREAT WHITE SHARK
Carcharodon carcharias

SAND TIGER SHARK
Carcharias taurus

GOBLIN SHARK
Mitsukurina owstoni

Carpetsharks, Bullhead Sharks, Frilled Sharks and Cow Sharks

Location

- Bullhead sharks, frilled sharks and cow sharks
- Bullhead sharks, frilled sharks, cow sharks and carpetsharks

WHALE SHARK FILTER FEEDING

Gill slits filter food from water.

Shark swallows filtered food.

Plankton, small fishes and water

Sharks

FISHES

CAMOUFLAGE

Some sharks have patterns on their skin that allow them to blend in with their environments. This helps conceal them from prey and predators.

Tasseled wobbegong has markings that make it resemble a rock.

Black-backed butterflyfish

Oval spot butterflyfish

442

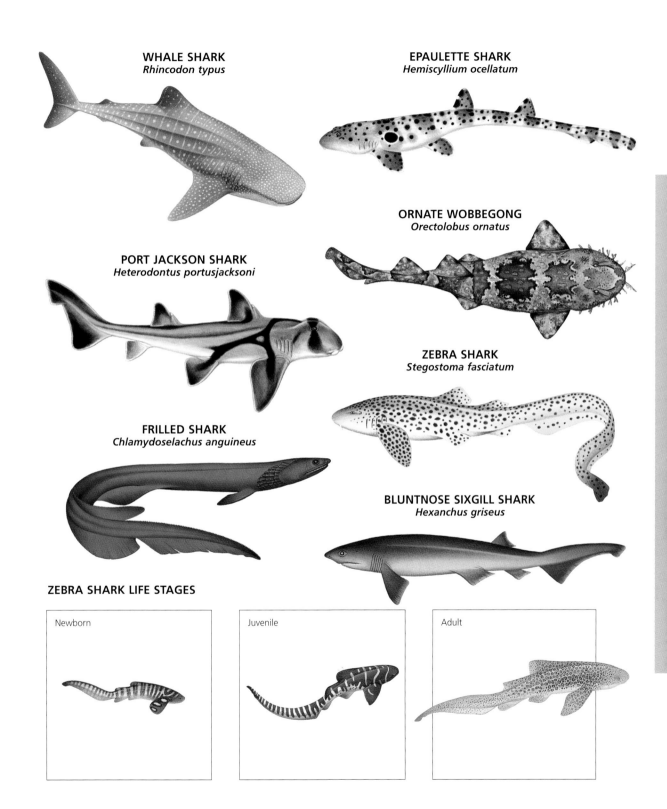

WHALE SHARK
Rhincodon typus

EPAULETTE SHARK
Hemiscyllium ocellatum

ORNATE WOBBEGONG
Orectolobus ornatus

PORT JACKSON SHARK
Heterodontus portusjacksoni

ZEBRA SHARK
Stegostoma fasciatum

FRILLED SHARK
Chlamydoselachus anguineus

BLUNTNOSE SIXGILL SHARK
Hexanchus griseus

ZEBRA SHARK LIFE STAGES

Newborn

Juvenile

Adult

Rays and Chimaeras

Location

- Rays and skates
- Chimaeras
- Rays, skates and chimaeras

FISH TAILS

Skate tail

Ray tail with sharp, poisonous spine

MAKING A MOVE

A shark's tail drives it forward and its pectoral fins help it swim upward or downward. Manta rays flap their pectoral fins to move through water. Stingrays use pectoral muscles to ripple along the sea floor.

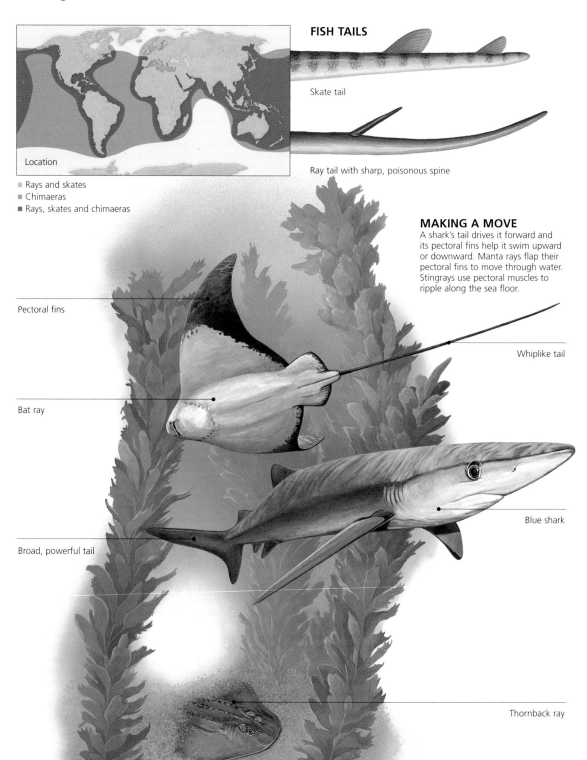

Pectoral fins

Bat ray

Whiplike tail

Blue shark

Broad, powerful tail

Thornback ray

Spookfish

Atlantic devil ray

Freshwater sawfish snout

Shortnose chimaera

Butterfly ray tooth

Skate

Elephantfish

Stingray

Guitarfish

Electric ray

Electric torpedo ray

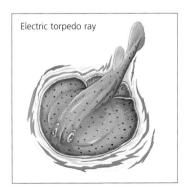

Mouth and nostrils of Atlantic torpedo ray

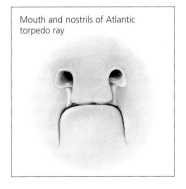

445

Rays and Chimaeras

BLUNT-NOSED CHIMAERA
Hydrolagus colliei

OCELLATED FRESHWATER STINGRAY
Potamotrygon motoro

SIXGILL STINGRAY
Hexatrygon bickelli

BLIND ELECTRIC RAY
Typhlonarke aysoni

BLUE SKATE
Notoraja sp.

PORT DAVEY SKATE
Raja sp.

HOW THEY SWIM
Birds can overcome gravity because they don't weigh much for their size. The same is true of cartilaginous fishes. Cartilage is lighter than bone, and this allows sharks and rays to slice through the water with minimum effort.

MANTA RAY FEEDING

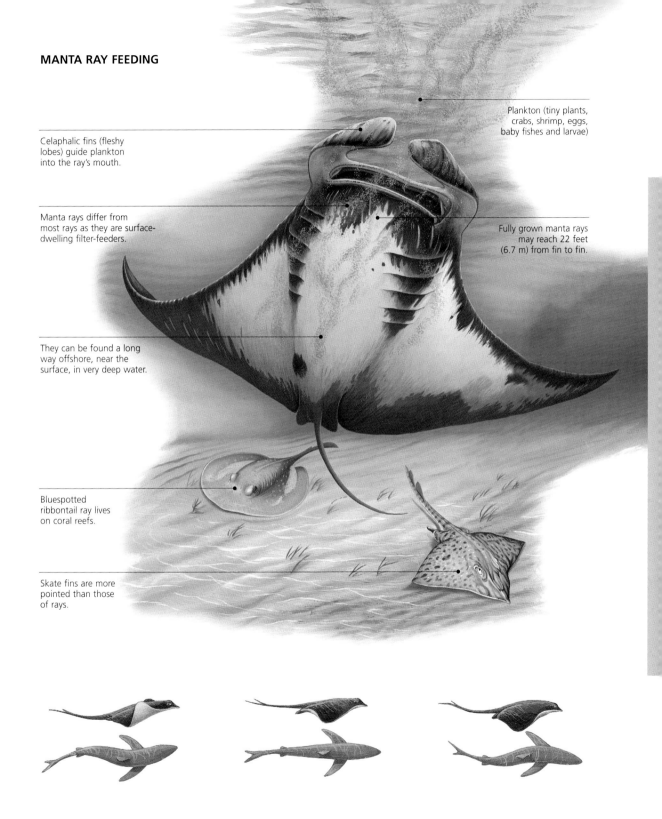

Celaphalic fins (fleshy lobes) guide plankton into the ray's mouth.

Manta rays differ from most rays as they are surface-dwelling filter-feeders.

They can be found a long way offshore, near the surface, in very deep water.

Bluespotted ribbontail ray lives on coral reefs.

Skate fins are more pointed than those of rays.

Plankton (tiny plants, crabs, shrimp, eggs, baby fishes and larvae)

Fully grown manta rays may reach 22 feet (6.7 m) from fin to fin.

Bony Fishes

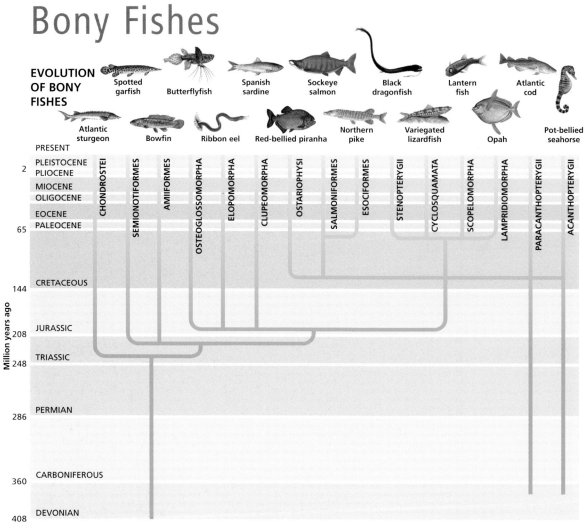

EVOLUTION OF BONY FISHES

Spotted garfish
Butterflyfish
Spanish sardine
Sockeye salmon
Black dragonfish
Lantern fish
Atlantic cod

Atlantic sturgeon
Bowfin
Ribbon eel
Red-bellied piranha
Northern pike
Variegated lizardfish
Opah
Pot-bellied seahorse

PRESENT
PLEISTOCENE
PLIOCENE
2
MIOCENE
OLIGOCENE
EOCENE
PALEOCENE
65
CRETACEOUS
144
JURASSIC
208
TRIASSIC
248
PERMIAN
286
CARBONIFEROUS
360
DEVONIAN
408

Million years ago

CHONDROSTEI
SEMIONOTIFORMES
AMIIFORMES
OSTEOGLOSSOMORPHA
ELOPOMORPHA
CLUPEOMORPHA
OSTARIOPHYSI
SALMONIFORMES
ESOCIFORMES
STENOPTERYGII
CYCLOSQUAMATA
SCOPELOMORPHA
LAMPRIDIOMORPHA
PARACANTHOPTERYGII
ACANTHOPTERYGII

Bony Fishes

FISHES

TROUT REPRODUCTION

Female brown trout lays eggs.

Eggs are fertilized by the male.

Hatchlings feed on orange yolk sac.

BLUE-FIN TUNA

TUNA SKELETON

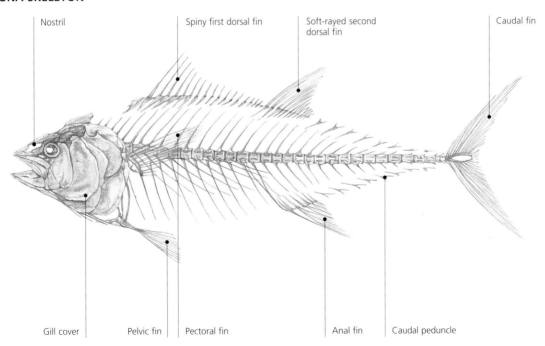

Nostril

Spiny first dorsal fin

Soft-rayed second
dorsal fin

Caudal fin

Gill cover

Pelvic fin

Pectoral fin

Anal fin

Caudal peduncle

Lungfishes and Coelacanth

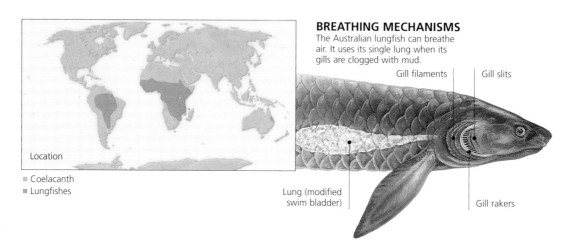

BREATHING MECHANISMS
The Australian lungfish can breathe air. It uses its single lung when its gills are clogged with mud.

Gill filaments

Gill slits

Location

- Coelacanth
- Lungfishes

Lung (modified swim bladder)

Gill rakers

SOUTH AMERICAN LUNGFISH
Lepidosiren paradoxa

AUSTRALIAN LUNGFISH
Neoceratodus forsteri

AFRICAN LUNGFISH
Protopterus dolloi

COELACANTH
Latimeria chalumnae

CRAWLING IN WATER
The African lungfish can swim
and crawl. When crawling, it
raises its body and propels itself
forward with its thin, paired fins.

Bichirs

Bony Fishes

FISHES

Location

■ Bichirs and relatives

MOTTLED BICHIR
Polypterus weeksi

ATLANTIC STURGEON
Acipenser oxyrinchus

BOWFIN
Amia calva

SPOTTED GAR
Lepisosteus oculatus

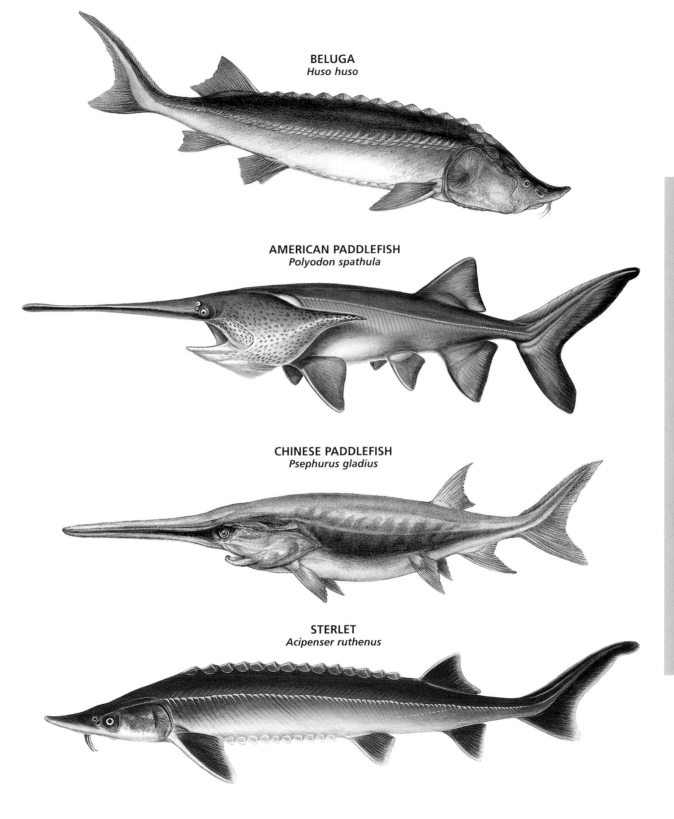

BELUGA
Huso huso

AMERICAN PADDLEFISH
Polyodon spathula

CHINESE PADDLEFISH
Psephurus gladius

STERLET
Acipenser ruthenus

Bonytongues

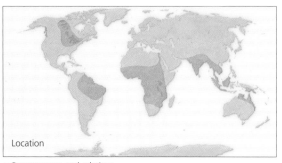

Location

■ Bonytongues and relatives

DORSAL-BAND WHALE
Petrocephalus simus

BUTTERFLYFISH
Pantodon buchholzi

ARAWANA
Osteoglossum bicirrhosum

ELEPHANTNOSE
Gnathonemus petersii

ASIATIC FEATHERBACK
Chitala chitala

MOONEYE
Hiodon tergisus

Eels

Location

■ Eels and relatives

Silver or mature

AMERICAN EEL
Anguilla rostrata

JUVENILE GARDEN EELS
Being able to burrow tail-first into the shell
bottom of tropical seas, garden eels are adapted
to life firmly anchored into the sea floor. They
emerge from their burrows to feed on plankton.

Yellow or juvenile

**TESSELLATED
MORAY EEL**
Gymnothorax favagineus

RIBBON EEL
Rhinomuraena quaesita

UMBRELLA MOUTH GULPER EEL
Eurypharynx pelecanoides

Luminous organ emits a
continuous pink glow
punctuated by red flashes.

Jaws nearly one-fourth
the total body length

Tiny eye

Tapering,
whiplike tail

Hatchet fish

DEEPSEA DWELLERS
Gulper eels live in a dark, slow-moving,
sparsely inhabited world over 6,500 feet
(1,980 m) below the surface of the sea.
Their hunt for prey is aided by a huge
mouth, prickly teeth, and a luminous
organ at the end of a long tail.

457

Sardines

Location

■ Sardines

ATLANTIC THREAD HERRING
Opisthonema oglinum

PERUVIAN ANCHOVETA
Engraulis ringens

AMERICAN GIZZARD SHAD
Dorosoma cepedianum

CASPIAN SPRAT
Clupeonella cultriventris

EUROPEAN PILCHARD
Sardina pilchardus

AMERICAN SHAD
Alosa sapidissima

BLACKFIN WOLF-HERRING
Chirocentrus dorab

SPANISH SARDINE
Sardinella aurita

TYPES OF SARDINES

Anchovies

Pacific herring

Sardines

Catfishes, Carps and Characins

Location

- Charachins
- Catfishes and carps
- Charachins, catfishes and carps

BITTERLING
Rhodeus sericeus

Breeding habits
Female bitterling lays eggs inside a mussel. The male sheds sperm over the mussel's syphon. As the mussel breathes, the sperm is drawn into the cavity to fertilize the eggs.

MILKFISH
Chanos chanos

SLIMY LOACH
Acantophthalmus myersi

EGG INCUBATION
The spotted African squeaker begins life in the mouth of another species of fish. A female cichlid picks up the eggs and broods them in her mouth.

A LEAP FOR LIFE
Splash tetra leap out of the water and spawn on the underside of objects such as leaves. The male splashes water on the eggs with his tail until they hatch.

ELECTRIC CATFISH
Malapterurus electricus

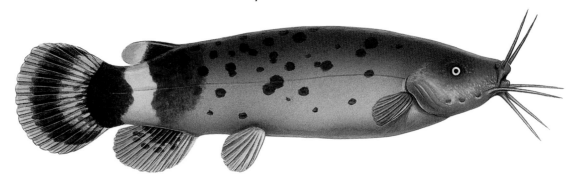

SADDLED HILLSTREAM LOACH
Homaloptera orthogoniata

TRICOLOR SHARKMINNOW
Balantiocheilos melanopterus

Catfishes, Carps and Characins

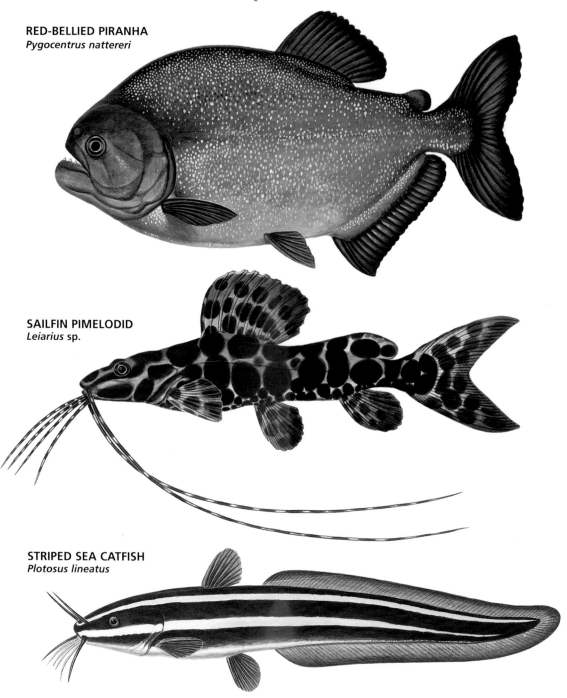

RED-BELLIED PIRANHA
Pygocentrus nattereri

SAILFIN PIMELODID
Leiarius sp.

STRIPED SEA CATFISH
Plotosus lineatus

MARBLED HATCHETFISH
Carnegiella strigata

STRIPED HEADSTANDER
Anostomus anostomus

ZEBRA DANIO
Danio rerio

Salmons

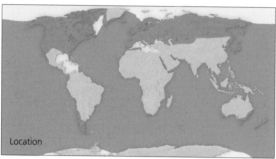

Location

■ Salmons

PIKE HEAD SKELETON

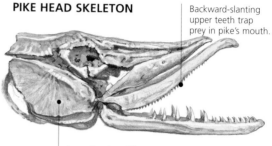

Backward-slanting upper teeth trap prey in pike's mouth.

Large head enables pike to swallow prey almost half its size.

SOCKEYE SALMON LIFE STAGES

Parr lives in rivers and small lakes.

It enters the sea as a smolt.

Smolt matures into a sea-going adult.

ARTHUR'S PARAGALAXIAS
Paragalaxias mesotes

CISCO
Coregonus artedi

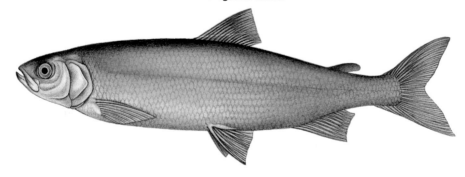

SOCKEYE SALMON LIFECYCLE

Summer–Autumn
Female buries several thousand fertilized eggs in a shallow nest dug in the gravel of a river bed.

Summer–Autumn
During spawning the male turns a brilliant red and develops a long, hooked jaw. After spawning, the salmon soon die.

■ Location

Winter–Spring
Eggs take several months to hatch. Hatchlings remain in the gravel until the attached yolk has been consumed.

Spring
Salmon fry wriggle out into the stream, inhabiting a lake for a year before beginning the journey out to sea.

■ Migratory path

Spring
The sea-bound salmon, known as smolts, face many predators such as other fishes, birds and mammals.

Spring–Summer
Returning to their birth river, some are eaten by brown bears. The survivors begin to show dramatic spawning colors.

Spring–Summer
After years at sea, salmon approach sexual maturity and are ready to face the rigors of the spawning migration.

Salmons

RAINBOW TROUT
Oncorhynchus mykiss

GOLDEN TROUT
Oncorhynchus aguabonita

LAKE TROUT
Salvelinus namaycush

GRAYLING
Thymallus thymallus

NORTHERN PIKE
Esox lucius

ALASKA BLACKFISH
Dallia pectoralis

SOCKEYE SALMON
Oncorhynchus nerka

ATLANTIC SALMON
Salmo salar

Dragonfishes and Lanternfishes

DRAGONFISH
Family *Stomiidae*

Location

■ Dragonfishes, lanternfishes and lizardfishes

BRISTLEMOUTH
Family *Gonostomatidae*

CROSS-TOOTHED PERCH
Order *Stomiiformes*

LANTERNFISH
Family *Myctophidae*

VARIEGATED LIZARDFISH
Synodus variegatus

VIPERFISH
Family *Stomiidae*

PACIFIC VIPERFISH
Chauliodus macouni

HATCHETFISH
Family *Sternoptychidae*

TRANSFORMATION
The larval forms of the black dragonfish have elongated stalks supporting their eyes. These stalks are absorbed as the fishes grow, until the eyes retreat into sockets.

Adult female

Larval form

BLACK DRAGONFISH
Idiacanthus fasciola

Cods and Anglerfishes

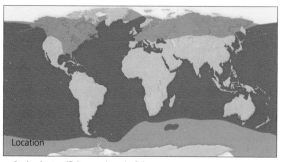

Location

- Cuskeels, toadfishes and anglerfishes
- Cods and troutperches
- Cuskeels, toadfishes, anglerfishes, cods and troutperches

ATLANTIC COD
Gadus morhua

PIRATE PERCH
Aphredoderus sayanus

SPLENDID TOADFISH
Sanopus splendidus

Bony Fishes

FISHES

470

PAXTON'S WHIPNOSE ANGLER
Gigantactis paxtoni

CIRCUMPOLAR BURBOT
Lota lota

PARASITIC LIFESTYLE

Male anglerfishes are tiny compared to females, with toothless jaws and no baits to lure prey. Males sometimes attach themselves to females and live as parasites.

CAMOUFLAGE TECHNIQUES

Bottom-dwelling anglerfishes blend in with their environment to snare prey. Frogfishes can change color and hide among coral.

Roughjaw frogfish

Sargassum fish

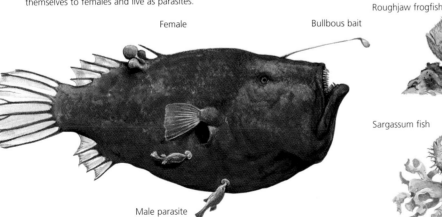

Female

Bullbous bait

Male parasite

Spiny-rayed Fishes

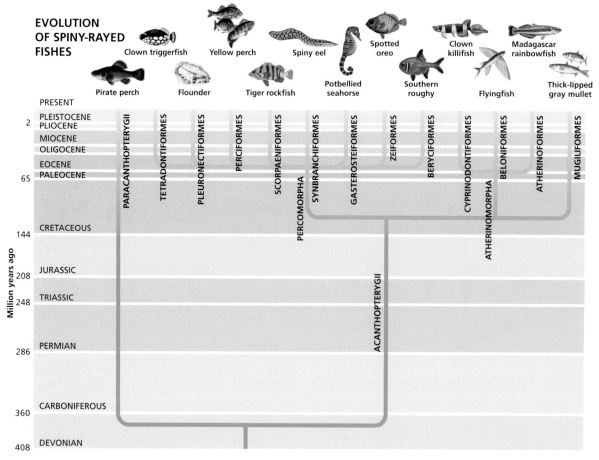

EVOLUTION OF SPINY-RAYED FISHES

Clown triggerfish
Yellow perch
Spiny eel
Spotted oreo
Clown killifish
Madagascar rainbowfish
Pirate perch
Flounder
Tiger rockfish
Potbellied seahorse
Southern roughy
Flyingfish
Thick-lipped gray mullet

Bony Fishes

FISHES

Million years ago

PRESENT
PLEISTOCENE
PLIOCENE
2
MIOCENE
OLIGOCENE
EOCENE
PALEOCENE
65
CRETACEOUS
144
JURASSIC
208
TRIASSIC
248
PERMIAN
286
CARBONIFEROUS
360
DEVONIAN
408

PARACANTHOPTERYGII
TETRADONTIFORMES
PLEURONECTIFORMES
PERCIFORMES
SCORPAENIFORMES
PERCOMORPHA
SYNBRANCHIFORMES
GASTEROSTEIFORMES
ZEIFORMES
BERYCIFORMES
CYPRINODONTIFORMES
ATHERINOMORPHA
BELONIFORMES
ATHERINOFORMES
MUGILIFORMES
ACANTHOPTERYGII

STICKLEBACK COURTSHIP

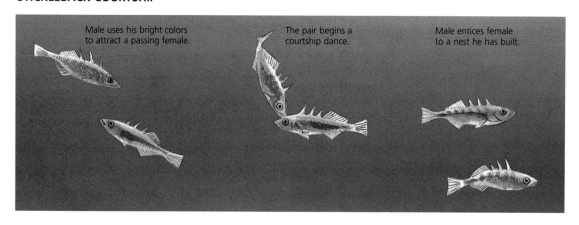

Male uses his bright colors to attract a passing female.

The pair begins a courtship dance.

Male entices female to a nest he has built.

AUSTRALIAN RAINBOWFISH

RAINBOWFISH SKELETON

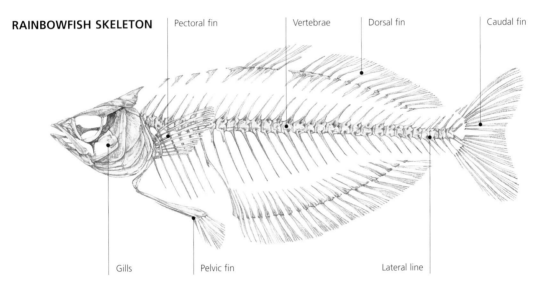

Pectoral fin

Vertebrae

Dorsal fin

Caudal fin

Gills

Pelvic fin

Lateral line

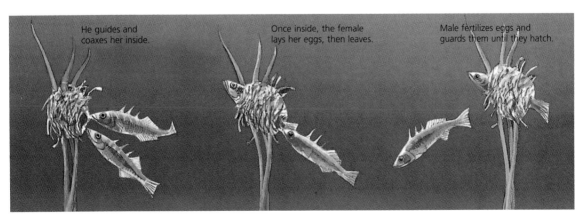

He guides and
coaxes her inside.

Once inside, the female
lays her eggs, then leaves.

Male fertilizes eggs and
guards them until they hatch.

Clingfishes, Flyingfishes, Killifishes, Ricefishes and Silversides

Location

■ Killifishes, ricefishes and silversides
■ Clingfishes and flyingfishes
■ Killifishes, ricefishes, silversides, clingfishes and flyingfishes

MANDARIN FISH
Synchiropus splendidus

Bony Fishes
FISHES

SWORDTAIL
Xiphophorus helleri

Changing sexes
The male swordtail is smaller than the female, and has a swordlike extension to its tail. Females sometimes turn into males, even after giving birth.

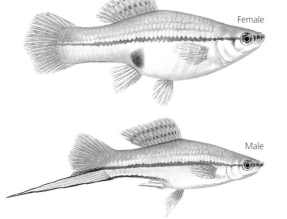

Female

Male

URCHIN CLINGFISH
Diademichthys lineatus

FLYINGFISH
Family *Exocoetidae*

MALAYAN HALFBEAK
Dermogenys pusilla

CLOWN KILLIFISH
Aplocheilus annulatus

JAVANESE RICEFISH
Oryzias javaniaus

STEEL-BLUE KILLIFISH
Aphyosemion gardneri

MADAGASCAR RAINBOWFISH
Bedotia geayi

CUATRO OJOS
Anableps anableps

Oarfishes, Squirrelfishes and Dories

Location

- Oarfishes and squirrelfishes
- Oarfishes, squirrelfishes and dories

FLASHING ON AND OFF

Flashlight fishes have light organs which contain luminous bacteria. To hide from predators, the fish covers the light organ with a type of eyelid called a melanphore.

Exposed

Light organ

Hidden

Melanphore

FLASHLIGHT FISH
Photoblepharon palpebratus

OARFISH
Regalecus glesne

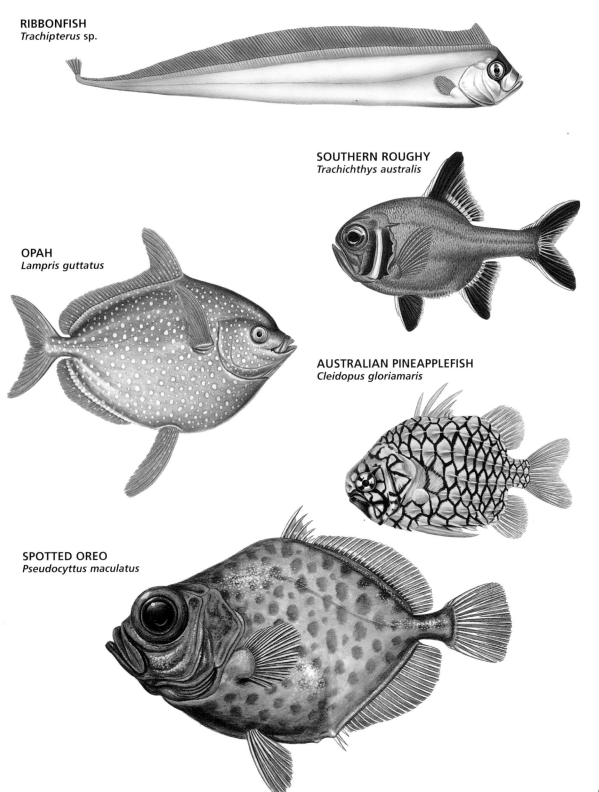

RIBBONFISH
Trachipterus sp.

SOUTHERN ROUGHY
Trachichthys australis

OPAH
Lampris guttatus

AUSTRALIAN PINEAPPLEFISH
Cleidopus gloriamaris

SPOTTED OREO
Pseudocyttus maculatus

Pipefishes, Swampeels and Scorpionfishes

Location

- Swampeels
- Pipefishes and scorpionfishes
- Swampeels, pipefishes and scorpionfishes

POT-BELLIED SEAHORSE
Hippocampus abdominalis

BANDED PIPEFISH
Doryrhamphus dactyliophorus

WHITE SPOTTED SPINY EEL
Mastacembelus armatus

CARIBBEAN TRUMPETFISH
Aulostomus maculatus

SEAHORSE REPRODUCTION

The female lays eggs in the male's marsupial-like pouch, leaving them in his care. The eggs are incubated in his pouch until they hatch.

LUMPSUCKER
Cyclopterus lumpus

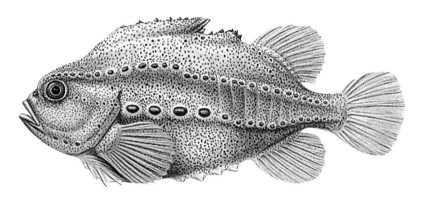

TIGER ROCKFISH
Sebastes nigrocinctus

COCKATOO WASPFISH
Ablabys taenionotus

Perches, Groupers and Seabasses

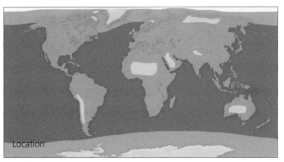

Location

- Perches
- Perches, groupers and seabasses

SPLENDID LICORICE GOURAMI
Parosphromenus dreissneri

SIXLINE SOAPFISH
Grammistes sexlineatus

PURPLEQUEEN
Pseudanthis tuka

COMMON DOLPHINFISH
Coryphaena hippurus

REMORA
Echeneis naucrates

GIANT GROUPER
Epinephelus lanceolatus

SOUTH-EAST ASIA PIKEHEAD
Luciocephalus pulcher

MEYER'S BUTTERFLYFISH
Chaetodon meyeri

MOORISH IDOL
Zanclus cornutus

Cichlids, Damselfishes, Wrasses, Parrotfishes and Blennies

Location

- Cichlids, damselfishes, wrasses and parrotfishes
- Cichlids, damselfishes, wrasses, parrotfishes and blennies

MUDSKIPPERS

The male protects eggs laid by the female by wrapping his body around them.

CHANGING COLOR AND SEX

As they mature, highfin parrotfishes (*Scarus altipinnis*) travel through three color phases, which also indicate changes to gender.

Juvenile: usually asexual female

Initial phase: usually female

Terminal phase: always a mature male

HARLEQUIN TUSKFISH
Choerodon fasciatus

STRIPED JULIE
Julidochromis regani

BLACK-HEADED BLENNY
Lipophrys nigriceps

OSCAR
Astronotus ocellatus

SPINECHEEK ANEMONEFISH
Premnas biaculeatus

RAINBOW CALE
Odax acroptilus

LAKE MALAWI ZEBRA CICHLID
Pseudotropheus zebra

483

Gobies, Flatfishes and Triggerfishes

Location

- Gobies
- Flatfishes and triggerfishes
- Gobies, flatfishes and triggerfishes

OCEAN SUNFISH
Mola mola

BLUEBANDED GOBY
Lythrypnus dalli

DORIA'S BUMBLEBEE GOBY
Brachygobius doriae

DECORATED FIRE-GOBY
Netmateleotris decora

RIGHT-EYED FLATFISH

The left eye of right-eyed flatfishes (flounders) moves toward the right eye. The front of the skull twists to bring the jaws sideways.

Larva—normal eye position

Left eye moves to top of head.

Adult—both eyes on right side

PEACOCK FLOUNDER
Bothus lunatus

PUFFERFISH

Pufferfishes inflate themselves into a spiny globe when threatened.

Relaxed

Threatened

SURVIVAL TECHNIQUES

The mimic filefish (*Paraluteres prionurus*) has evolved to look like the toxic blacksaddled puffer (*Canthigaster valentini*).

Blacksaddled puffer
(venomous)

Mimic filefish
(non-venomous)

FIGURE-EIGHT PUFFER
Tetraodon biocellatus

CLOWN TRIGGERFISH
Balistoides conspicillum

485

Classifying Invertebrates

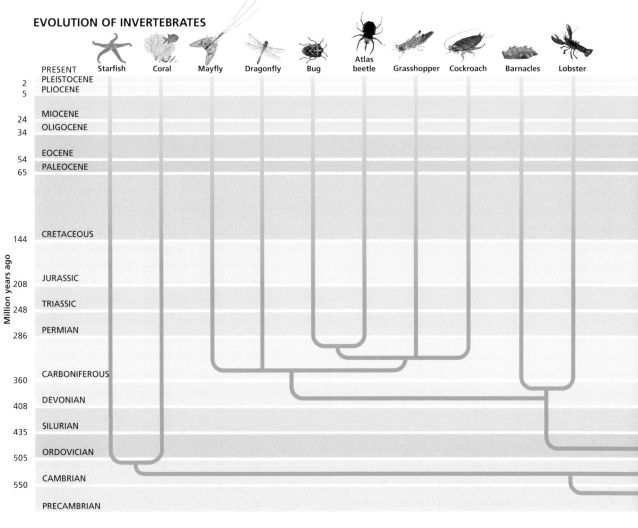

EVOLUTION OF INVERTEBRATES

Million years ago		
PRESENT		
2	PLEISTOCENE	
5	PLIOCENE	
24	MIOCENE	
34	OLIGOCENE	
54	EOCENE	
65	PALEOCENE	
144	CRETACEOUS	
208	JURASSIC	
248	TRIASSIC	
286	PERMIAN	
360	CARBONIFEROUS	
408	DEVONIAN	
435	SILURIAN	
505	ORDOVICIAN	
550	CAMBRIAN	
	PRECAMBRIAN	

Starfish   Coral   Mayfly   Dragonfly   Bug   Atlas beetle   Grasshopper   Cockroach   Barnacles   Lobster

Introducing Invertebrates

INVERTEBRATES

Sea anemones with clownfishes

Crab

Mollusk shells

490

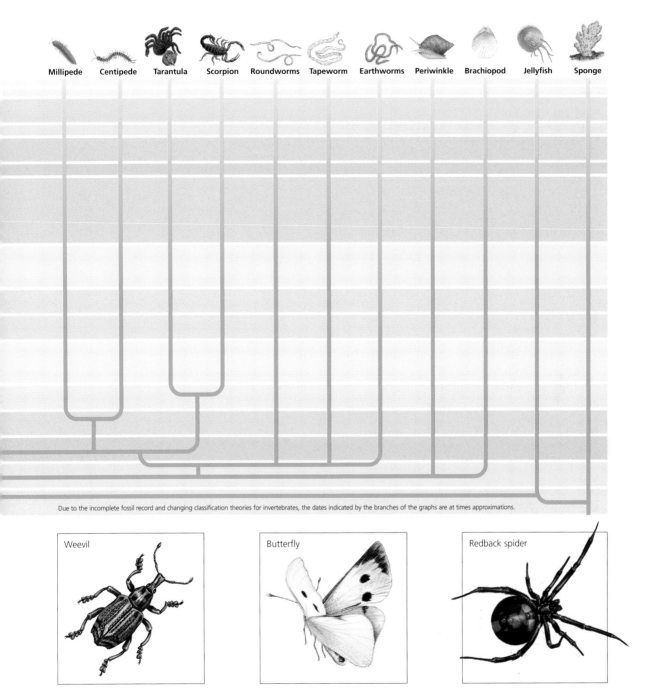

Millipede   Centipede   Tarantula   Scorpion   Roundworms   Tapeworm   Earthworms   Periwinkle   Brachiopod   Jellyfish   Sponge

Due to the incomplete fossil record and changing classification theories for invertebrates, the dates indicated by the branches of the graphs are at times approximations.

Weevil

Butterfly

Redback spider

491

Invertebrate Chordates, Sponges and Cnidarians

CORAL REEF ECOSYSTEM

Corals thrive in the shallow waters of the tropical Pacific and Indian oceans. Thousands of individual coral polyps, each protected by a limestone exoskeleton, group together to form living colonies; these in turn form coral reefs.

The crown-of-thorns starfish attacks and eats corals by disgorging its stomach over a colony, then absorbing the liquefied tissues.

Coral reefs provide shelter, food and breeding territory for thousands of species of plants and animals, such as fish, sharks and turtles.

Within certain coral communities, individual polyps have specialized functions—feeding, breeding or defense.

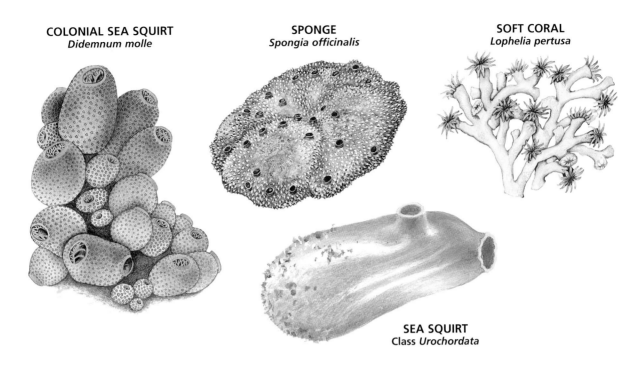

COLONIAL SEA SQUIRT
Didemnum molle

SPONGE
Spongia officinalis

SOFT CORAL
Lophelia pertusa

SEA SQUIRT
Class *Urochordata*

Sea anemone

Sea sponge

Portuguese man o'war

Inside a coral polyp

Jellyfish

Corals with fish

Mollusks

MUSSEL LIFECYCLE

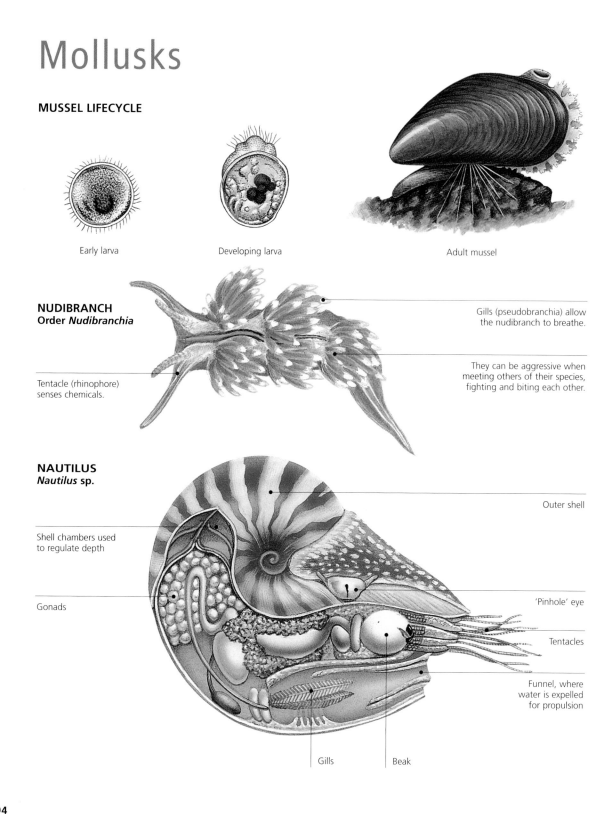

Early larva

Developing larva

Adult mussel

NUDIBRANCH
Order *Nudibranchia*

Tentacle (rhinophore) senses chemicals.

Gills (pseudobranchia) allow the nudibranch to breathe.

They can be aggressive when meeting others of their species, fighting and biting each other.

NAUTILUS
***Nautilus* sp.**

Shell chambers used to regulate depth

Gonads

Outer shell

'Pinhole' eye

Tentacles

Funnel, where water is expelled for propulsion

Gills

Beak

Common garden snail

Periwinkle (marine snail)

Periwinkle displaying muscular foot

Squid

Oyster shells

Oyster with pearl—layers of mucus surrounding foreign body

Brachiopod shells

Limpet displaying muscular foot

Mussels

Bivalve shells

Mud whelk

Common garden slug

Squids and Octopuses

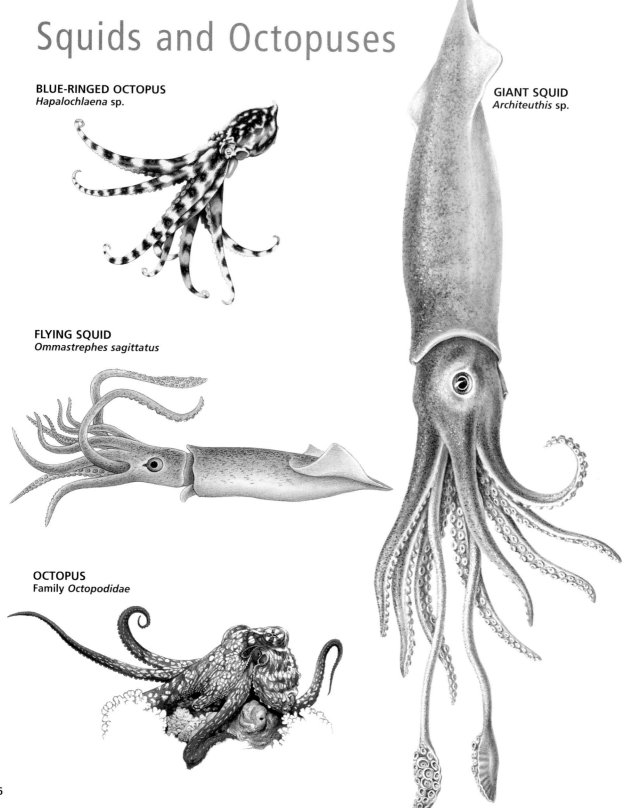

BLUE-RINGED OCTOPUS
Hapalochlaena sp.

GIANT SQUID
Architeuthis sp.

FLYING SQUID
Ommastrephes sagittatus

OCTOPUS
Family *Octopodidae*

CLASH OF THE TITANS

It is unknown exactly how giant squid are hunted by sperm whales, and until only recently, the battle scars worn by sperm whales, together with partially digested remains, were the only evidence that giant squid existed.

Powerful tail flukes enable the sperm whale to dive vertically into the ocean's depths.

Flipper

Tentacles equipped with suction pads, called clubs, are used to capture prey.

Giant squid have never been observed in the wild, and little is known about their behavior.

They can exceed 30 feet (10 m) in length and are thought to live 600–3000 feet (200–1000 m) below sea level.

Eyes approximately the size of a human head are positioned on either side of the mouth.

Worms

PARASITIC WORMS

Hookworm eggs

Tapeworm head

WHIPWORMS
Trichuris sp.

MARINE TAPEWORM
Class *Cestoidea*

MARINE LEECH
Family *Piscicolidae*

ROUNDWORMS
Phylum *Nematoda*

TERRESTRIAL TAPEWORM
Class *Cestoidea*

EARTHWORMS
Class *Oligochaeta*

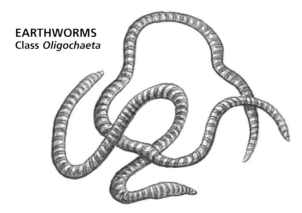

WORM EATING APPLE
While many worms are
parasitic (they survive
by living off their host),
others eat plant or animal
matter, such as fruit.

TUBE WORMS
Class *Pogonophora*

Tube worms such as these live in the depths of the ocean.

They are found in clusters near deep-sea vents, with one end attached to the ocean floor.

Simple red plume (breathing organ) contains hemoglobin.

Lacking eyes, stomach, or other major organs, the worms house bacteria that provide them with food.

Eelpout are predators of tube worms.

Echinoderms

STARFISH ANATOMY

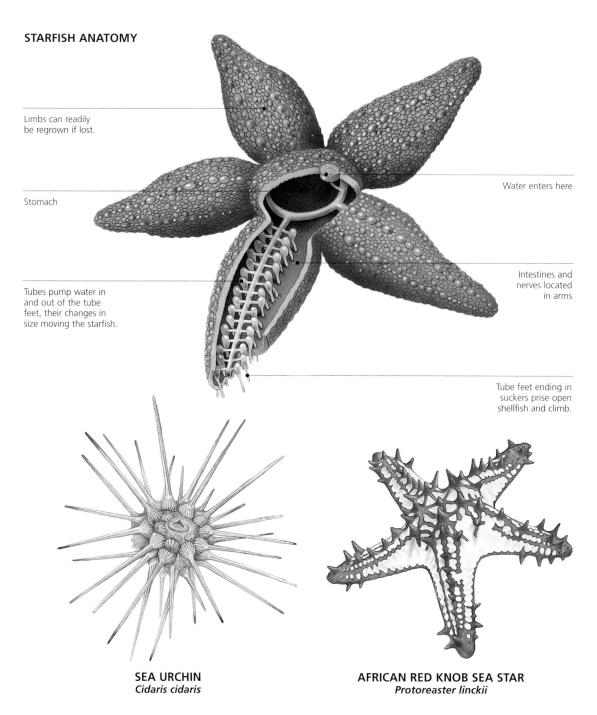

Limbs can readily
be regrown if lost.

Stomach

Tubes pump water in
and out of the tube
feet, their changes in
size moving the starfish.

Water enters here

Intestines and
nerves located
in arms

Tube feet ending in
suckers prise open
shellfish and climb.

SEA URCHIN
Cidaris cidaris

AFRICAN RED KNOB SEA STAR
Protoreaster linckii

SEA CUCUMBER
Stichopus chloronotus

CROWN-OF-THORNS STARFISH
Acanthaster planci

Nocturnal crown-of-thorns starfish feed on coral polyps when corals are most active.

Individual starfish can devour over 10 square miles (26 sq km) annually.

Sea star

Starfish

Baculogypsina

Calcarina

Starfish

Sea urchin

Arthropods

GREAT SURVIVORS
Arthropods are invertebrates that are protected by jointed exoskeletons. They are the largest and most successful group of living organisms on the planet (more than a million species), and include insects, spiders and crustaceans.

Grasshopper

Butterfly

Mantid

Cockchafer

Wasp

Ladybug

Mosquito

Cicada

Bug

Mayfly

Fly   Lacewing   Dragonfly

Spider

Tick

Hermit crab

Centipede

Arachnids

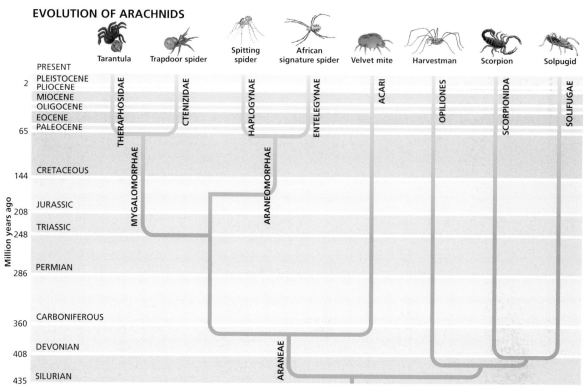

EVOLUTION OF ARACHNIDS

Tarantula | Trapdoor spider | Spitting spider | African signature spider | Velvet mite | Harvestman | Scorpion | Solpugid

PRESENT
PLEISTOCENE
PLIOCENE
MIOCENE
OLIGOCENE
EOCENE
PALEOCENE
CRETACEOUS
JURASSIC
TRIASSIC
PERMIAN
CARBONIFEROUS
DEVONIAN
SILURIAN

2
65
144
208
248
286
360
408
435

Million years ago

THERAPHOSIDAE
CTENIZIDAE
HAPLOGYNAE
ENTELEGYNAE
ACARI
OPILIONES
SCORPIONIDA
SOLIFUGAE

MYGALOMORPHAE
ARANEOMORPHAE
ARANEAE

ARACHNID ANATOMY

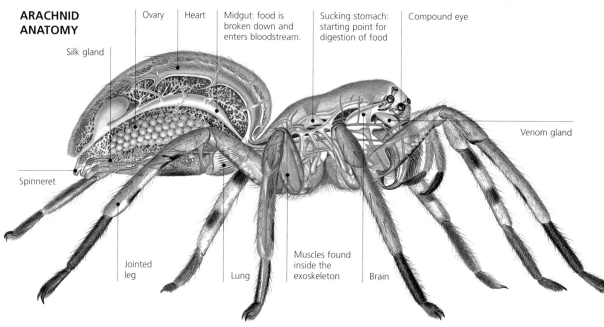

Silk gland

Ovary | Heart | Midgut: food is broken down and enters bloodstream. | Sucking stomach: starting point for digestion of food | Compound eye

Venom gland

Spinneret

Jointed leg

Lung

Muscles found inside the exoskeleton

Brain

DIVERSE WEAVERS

The orb-web spinning garden spider has two pairs of spinnerets, which produce two varieties of cocoon silk and several of web silk.

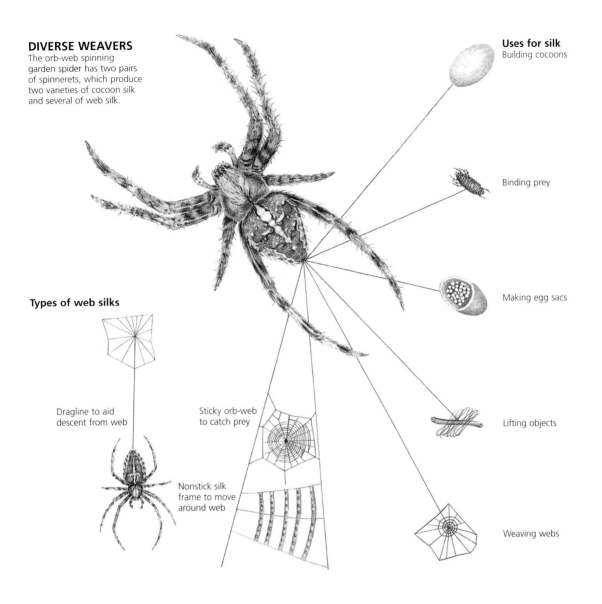

Uses for silk

Building cocoons

Binding prey

Making egg sacs

Lifting objects

Weaving webs

Types of web silks

Dragline to aid descent from web

Sticky orb-web to catch prey

Nonstick silk frame to move around web

SPIDER FACES

Crab spider

Woodlouse-eating spider

Ogre-faced spider

Huntsman spider

Spiders

WHITE LADY SPIDER
Family *Sparassidae*

PRIMITIVE SPIDER
Suborder *Labidognatha*

COMB-FOOTED SPIDER
Family *Theridiidae*

JUMPING SPIDER
Family *Salticidae*

TARANTULA
Family *Theraphosidae*

LYNX SPIDER
Family *Oxyopidae*

CRAB SPIDER
Family *Thomisidae*

COURTSHIP DISPLAYS

Wolf spiders

Jumping spiders

REPRODUCTION RITUALS

Nursery web spiders
Male presents wrapped prey
to female, before mating.

Crab spiders
Male fastens female to ground
with silk, then deposits sperm.

SPIDER JAWS

Mygalomorphic (hairy) spider
Fangs point downward

Arabeomorphic (typical) spider
Fangs hinge together sideways

SHEDDING THE EXOSKELETON

Hangs from web   Exoskeleton tears apart   Draws fragile legs out   Waits for exoskeleton to dry

Spiders

REDBACK SPIDER
Family *Theridiidae*

BOLAS SPIDER
Family *Araneidae*

NORTHERN BLACK WIDOW
Family *Theridiidae*

SPITTING SPIDER
Family *Scytodidae*

JUMPING SPIDERS
Family *Salticidae*

MOUSE SPIDER
Family *Theraphosidae*

TRAPDOOR SPIDER
Family *Ctenizidae*

FUNNEL-WEB SPIDER
Family *Agelenidae*

VIOLIN SPIDER
Family *Loxoscelidae*

FROM EGG TO ADULT

Huntsman egg sac with eggs

Huntsman spiderling

Juvenile huntsman

Adult huntsman

HUNTSMAN SPIDER
Family *Sparassidae*

BROWN WIDOW
Family *Theridiidae*

**BRAZILIAN
WANDERING SPIDER**
Family *Ctenidae*

RED WIDOW
Family *Theridiidae*

SOUTHERN BLACK WIDOW
Family *Theridiidae*

Venom
gland

FUNNEL-WEB FANGS

VENOM GLAND

Venom stored in the gland exits
through ducts into the fangs.

Sharp, downward-
facing fang

Venom drop

Funnel-web raises
its forelimbs and
body to strike.

Hindlimbs steady
the striking spider.

Orb-weaving Spiders

AFRICAN SIGNATURE SPIDER
Family *Araneidae*

BUILDING WEBS

Lace-sheet web

Triangle web

Hammock web

Scaffold web

510

MARBLED ORB WEAVER
Family *Araneidae*

SIGNATURE SPIDERS
Family *Araneidae*

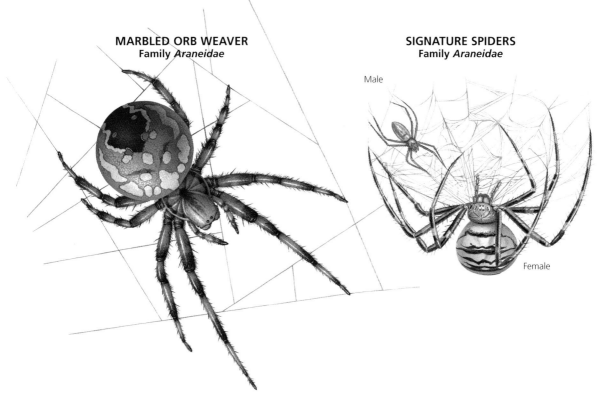

Male

Female

GOLDEN SILK SPIDER
Nephila sp.

SPINY ORB WEAVER
Family *Araneidae*

LONG-JAWED ORB WEAVER
Family *Araneidae*

CURVED SPINY SPIDER
Gasteracantha sp.

Scorpions, Mites and Ticks

HARVESTMAN
Family *Leiobunidae*

SOLPUGID
Family *Solpugidae*

HARD TICK
Family *Ixodidae*

SCORPION
Family *Buthidae*

VELVET MITE
Family *Trombidiidae*

Arthropods

INVERTEBRATES

SCORPION ANATOMY

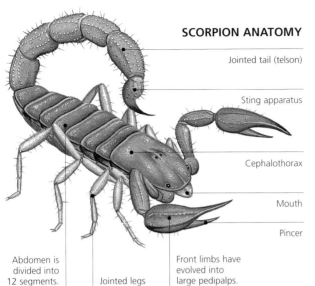

Jointed tail (telson)

Sting apparatus

Cephalothorax

Mouth

Pincer

Abdomen is divided into 12 segments.

Jointed legs

Front limbs have evolved into large pedipalps.

STING IN THE TAIL

Muscle moves position of sting apparatus.

Muscle squeezes poison gland to emit venom.

Stinger

SCORPION COURTSHIP DISPLAY

Some species of scorpions perform elaborate courtship displays, which involve joining pincers and 'dancing' around each other before mating.

Crustaceans, Silverfishes, Centipedes and Millipedes

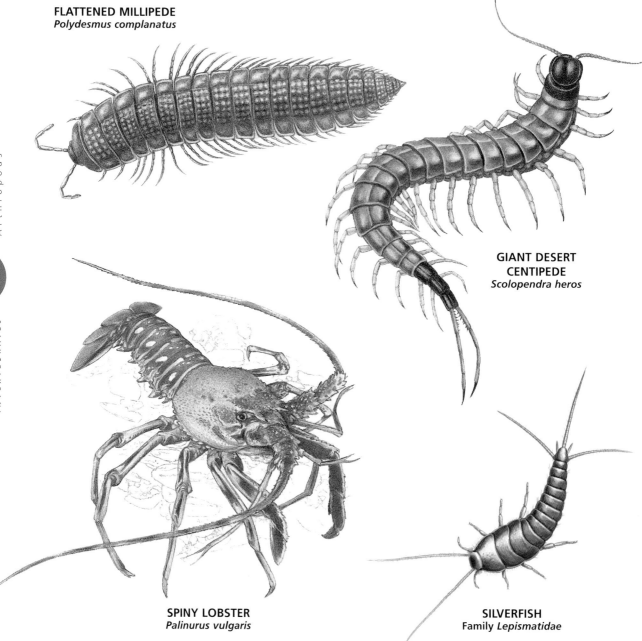

FLATTENED MILLIPEDE
Polydesmus complanatus

GIANT DESERT CENTIPEDE
Scolopendra heros

SPINY LOBSTER
Palinurus vulgaris

SILVERFISH
Family *Lepismatidae*

Lobster

Shrimp

Centipede

Crab

SPINY LOBSTER MIGRATION

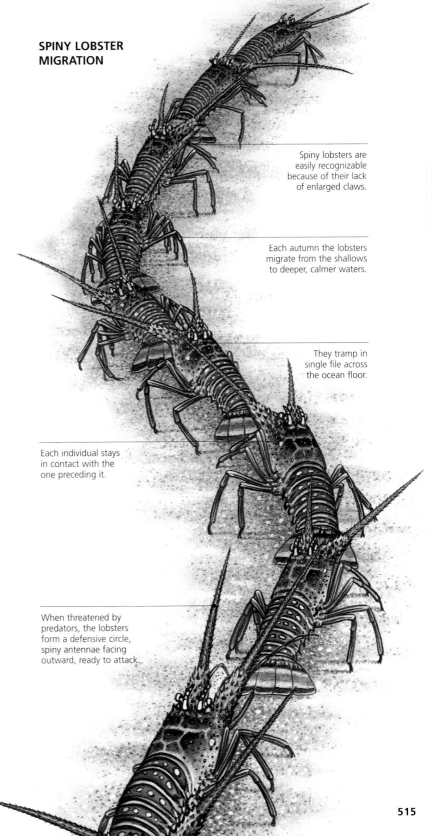

Spiny lobsters are easily recognizable because of their lack of enlarged claws.

Each autumn the lobsters migrate from the shallows to deeper, calmer waters.

They tramp in single file across the ocean floor.

Each individual stays in contact with the one preceding it.

When threatened by predators, the lobsters form a defensive circle, spiny antennae facing outward, ready to attack.

Crustaceans, Silverfishes, Centipedes and Millipedes

KRILL
Family *Euphausiidae*

MILLIPEDE
Class *Diplopoda*

SHRIMPS
Order *Decapoda*

BANDED CORAL SHRIMP
Stenopus hispidus

PURPLE SHORE CRAB
Hemigrapsus nudus

HERMIT CRAB
Family *Diogenidae*

CRAB LIFECYCLE
Crabs reproduce by releasing their larvae into the ocean, where they undergo a number of larval forms before returning to tidal areas as adults.

Zoea larva

Megalopa larva

CRAB
Order *Decapoda*

Insects

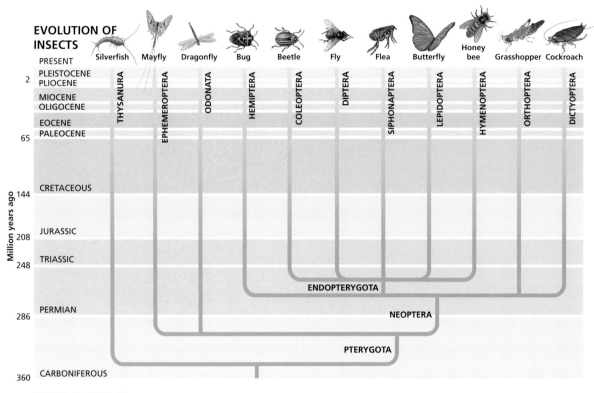

EVOLUTION OF INSECTS

Silverfish · Mayfly · Dragonfly · Bug · Beetle · Fly · Flea · Butterfly · Honey bee · Grasshopper · Cockroach

THYSANURA · EPHEMEROPTERA · ODONATA · HEMIPTERA · COLEOPTERA · DIPTERA · SIPHONAPTERA · LEPIDOPTERA · HYMENOPTERA · ORTHOPTERA · DICTYOPTERA

Million years ago

PRESENT
PLEISTOCENE
PLIOCENE — 2
MIOCENE
OLIGOCENE
EOCENE
PALEOCENE — 65
CRETACEOUS — 144
JURASSIC — 208
TRIASSIC — 248
PERMIAN — 286
CARBONIFEROUS — 360

ENDOPTERYGOTA

NEOPTERA

PTERYGOTA

Arthropods

INVERTEBRATES

TYPES OF WINGS

True fly

Mantid

Butterfly

Thrip

Wasp

Dragonfly

INSECT ANATOMY

Ocelli detect amount of light

Exoskeleton covered by tiny hairs

Crop

Wing made from same material as exoskeleton

Brain

Antenna

Mandible

Tongue

Compound eye

Thorax

Trachea

Nerve

Leg muscle

Abdomen

Heart

Midgut

Poison sac

Stinger

TYPES OF ANTENNAE

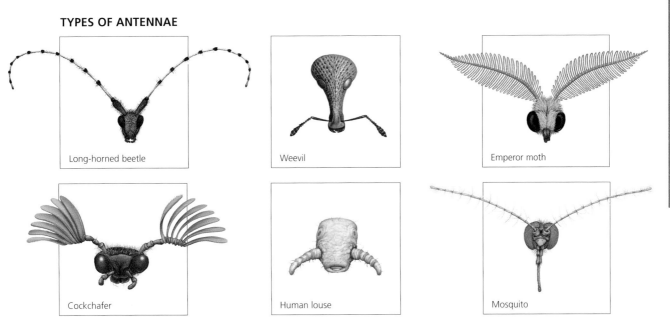

Long-horned beetle

Weevil

Emperor moth

Cockchafer

Human louse

Mosquito

Dragonflies, Mayflies and Mantids

DRAGONFLY ANATOMY

Ocelli monitor the horizon to keep dragonfly flying level.

Front wing

Rear wing beats at a different speed to front wing to increase stability.

Each lens of the compound eye sees things differently.

Antenna

Jointed leg

Hairs around mouth taste and guide food.

Arthropods

INVERTEBRATES

DRAGONFLY LIFECYCLE

Adult dragonflies are powerful fliers who hunt other insects.

New adult dragonfly basks in the sun while its wings extend.

Female lays her eggs in an aquatic plant below the water's surface.

Male (top) grabs the female with his tail and fertilizes her eggs.

Nymph climbs out of the water for its final molt into adulthood.

Nymph swallows air to burst itself out of its old skin.

Eggs may take several weeks to hatch as nymphs chew their way free.

Nymph catches tadpoles and worms with its hinged jaws.

Praying mantis forelimb

Dragonfly

Dragonflies in flight

Mantis

Damselfly nymph

Mayfly

PRAYING MANTIS
Family *Mantidae*

ORCHID MANTIS
Family *Hymenopodidae*

Cockroaches, Termites and Lice

DAMP-WOOD TERMITE
Family
Rhinotermitidae

FEMALE COCKROACH LAYING EGGS
As the eggs of many species of cockroaches are laid on the ground, they are covered in hard cases for protection.

Arthropods

INVERTEBRATES

Cockroach

Termite

Spinifex termite

Cockroach head displaying antennae

Louse

Fish louse

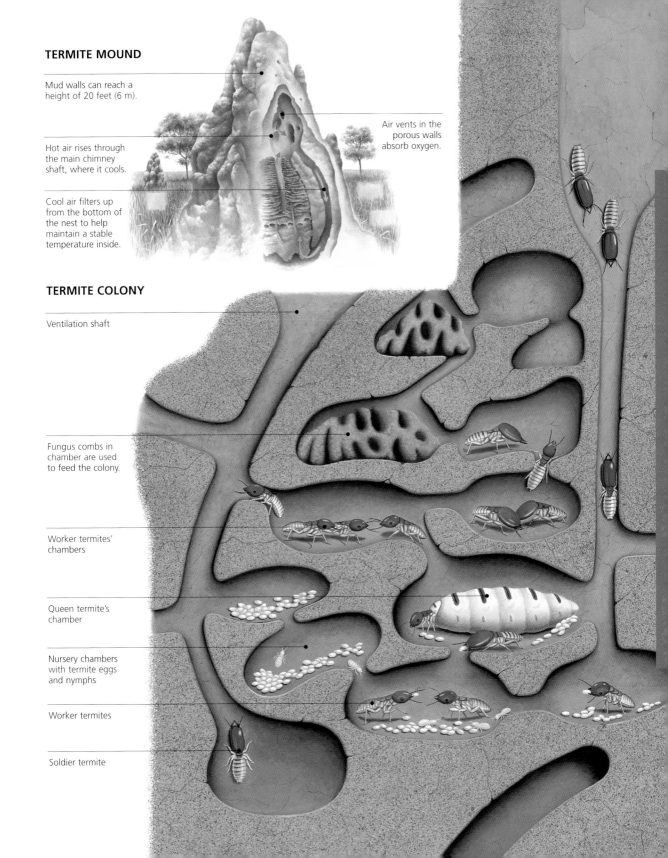

TERMITE MOUND

Mud walls can reach a height of 20 feet (6 m).

Hot air rises through the main chimney shaft, where it cools.

Cool air filters up from the bottom of the nest to help maintain a stable temperature inside.

Air vents in the porous walls absorb oxygen.

TERMITE COLONY

Ventilation shaft

Fungus combs in chamber are used to feed the colony.

Worker termites' chambers

Queen termite's chamber

Nursery chambers with termite eggs and nymphs

Worker termites

Soldier termite

Crickets and Grasshoppers

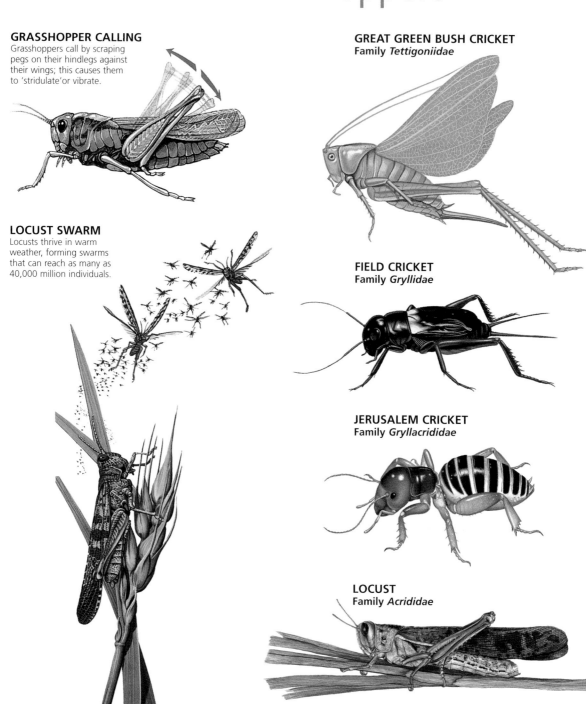

GRASSHOPPER CALLING
Grasshoppers call by scraping pegs on their hindlegs against their wings; this causes them to 'stridulate'or vibrate.

GREAT GREEN BUSH CRICKET
Family *Tettigoniidae*

LOCUST SWARM
Locusts thrive in warm weather, forming swarms that can reach as many as 40,000 million individuals.

FIELD CRICKET
Family *Gryllidae*

JERUSALEM CRICKET
Family *Gryllacrididae*

LOCUST
Family *Acrididae*

Arthropods

INVERTEBRATES

Male mole cricket in tunnel

Bush cricket

Koringkriek cricket leg

True katydid on leaf

Grasshopper spreading wings

Grasshopper

Adult earwig and nymphs feeding on branch

Stick insect

Grasshopper head

Nymph (top) and adult earwigs

Grasshopper leg

Nymph (top) and adult locusts

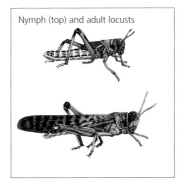

Bugs, Lacewings and Thrips

CICADA
Family *Cicadidae*

Anatomy of feeding
Bugs have distinctive mouthparts—
a beaklike rostrum surrounded by
four sharp stylets. The stylets
pierce the food source, then the
rostrum carries toxic saliva through
the puncture to partially digest the
meal before ingesting.

Head

Eye

Mouth

Limbs well-adapted
for climbing

Stylets

Limbs attached
to thorax

Rostrum folds
underneath the body
in some species.

Abdomen

**LEAFHOPPER
LIFECYCLE**

Leafhopper waits on
blade of grass for
exoskeleton to harden.

Leafhopper nymph
prepares to molt
its final juvenile
exoskeleton.

Newly adult
leafhopper pulls
itself free of its
old exoskeleton.

Lacewing at rest

Lacewing taking off

Lacewing in flight

Stink bug

True bug with bright warning coloring

Milkweed bug

Lacewing

Thrip

Water bug leg

Nymph (top) and adult elder bugs

Periodical cicada

Assassin bug head

Bugs, Lacewings and Thrips

PEANUT BUGS
Fulgora sp.

Out of sight
Their dull coloration mimics the patterns of the trees they rest on, camouflaging them from hungry birds.

Spotted
If spied, a peanut bug snaps open its wings to reveal two large, eyelike spots to try to startle the would-be predator away.

Water scorpion attacks a tadpole while hanging from pond weeds.

JESTER BUG
Family *Heteroptera*

Water boatman swims through the pond, using its legs as paddles.

Water striders use surface tension to 'walk' on the water's surface.

Many species of bugs spend larval or juvenile stages in the water.

Beetles

HERCULES BEETLES
Family *Dynastinae*
Rival male Hercules beetles grapple with their swordlike horns, battling for the right to mate with the female (right).

Diving beetle leg

Whirligig beetle swimming

Fire beetle

Diving beetle underwater

Darkling beetle

Jewel beetle

530

Colorado beetle

Flea beetle

Hercules beetle

Colorado beetle
on branch

Large chafer

Rhinoceros beetle

531

Beetles

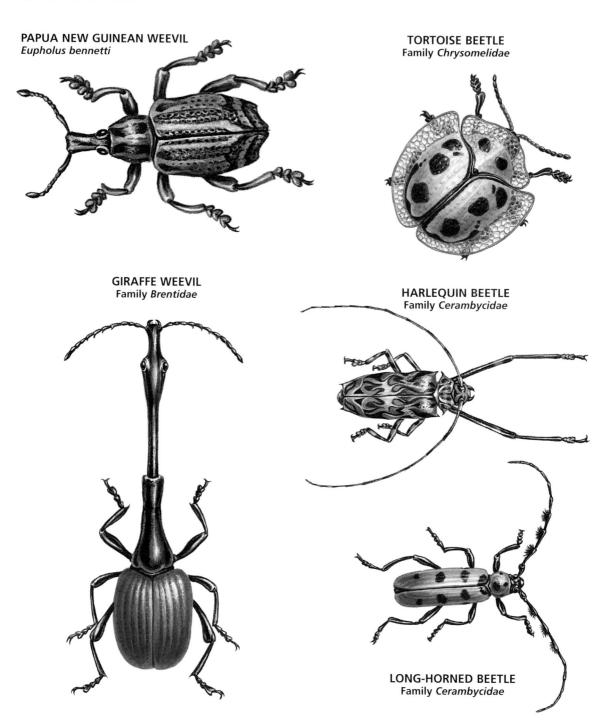

PAPUA NEW GUINEAN WEEVIL
Eupholus bennetti

TORTOISE BEETLE
Family *Chrysomelidae*

GIRAFFE WEEVIL
Family *Brentidae*

HARLEQUIN BEETLE
Family *Cerambycidae*

LONG-HORNED BEETLE
Family *Cerambycidae*

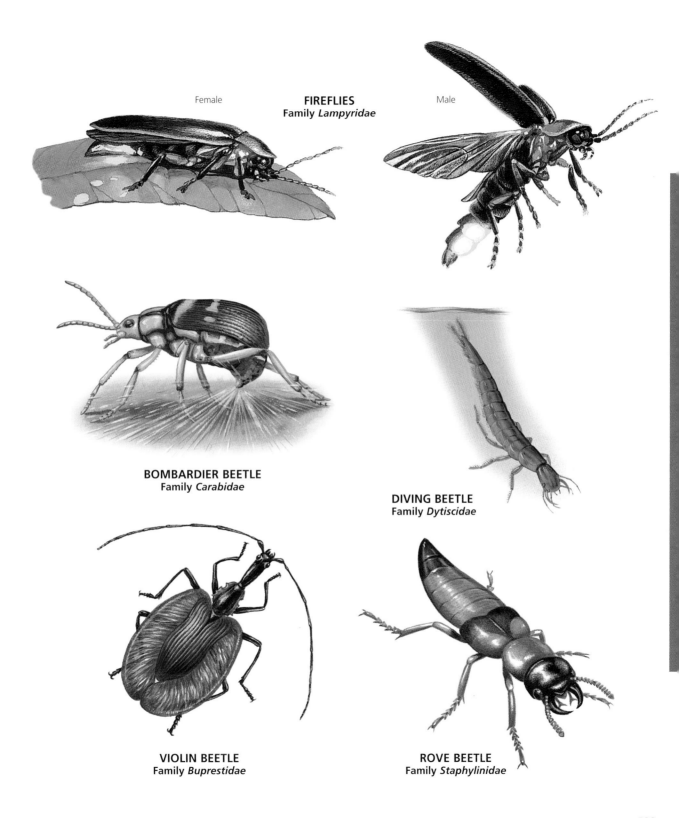

Female FIREFLIES
Family *Lampyridae* Male

BOMBARDIER BEETLE
Family *Carabidae*

DIVING BEETLE
Family *Dytiscidae*

VIOLIN BEETLE
Family *Buprestidae*

ROVE BEETLE
Family *Staphylinidae*

Ladybugs

Arthropods

TEN-SPOTTED LADYBUG
Family *Coccinellidae*

FIVE-SPOTTED LADYBUG
Family *Coccinellidae*

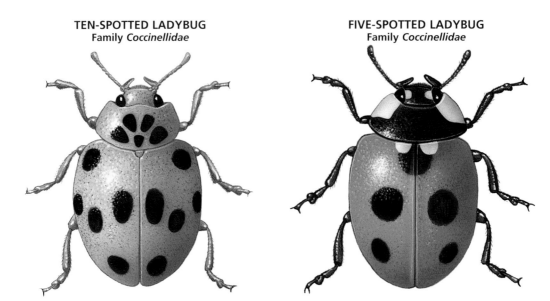

INVERTEBRATES

LADYBUG LAYING EGGS
The bright yellow eggs of the ladybug are laid on a leaf in clutches. The female chooses to lay her eggs close to a food source, such as an aphid colony.

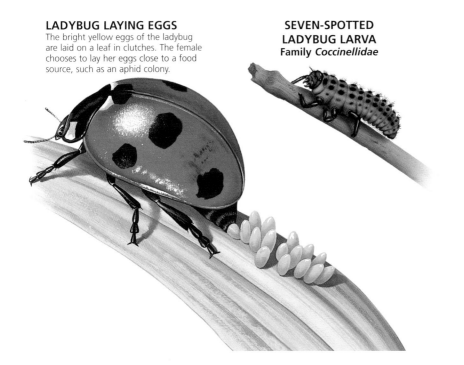

**SEVEN-SPOTTED
LADYBUG LARVA**
Family *Coccinellidae*

LADYBUG LIFE STAGES

Egg
Hatches at 3–5 days

Larva
Pupates at 2–3 weeks

Pupa
Changes after 7–10 days

Adult
Lives for more than one year

LADYBUGS EATING APHIDS

Slow-moving aphids are the main diet of ladybugs, who are the gardener's friend. Some species can eat more than 100 aphids per day.

IN FLIGHT

The hardened front wings of beetles cannot beat up and down. Instead, they provide uplift while the back wings propel the ladybug forward.

TAKING FLIGHT

Back wings are normally tucked away under the brightly colored, hard front wings.

To prepare for flight, the front wings are opened and back wings unfolded.

Front wings swing up and out, back wings beat rapidly and the ladybug takes off.

535

Flies, Fleas and Mosquitoes

DEER FLY
Family *Tabanidae*

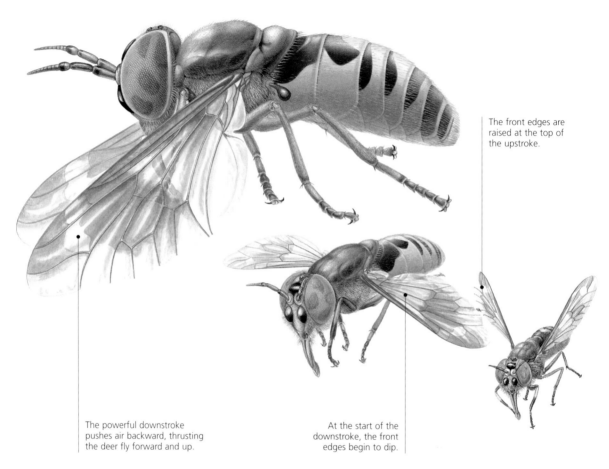

The front edges are raised at the top of the upstroke.

The powerful downstroke pushes air backward, thrusting the deer fly forward and up.

At the start of the downstroke, the front edges begin to dip.

CADDIS FLY LIFE STAGES

Egg

Larva

Pupa

Adult

Fly leg with sticky foot pads

Horsefly

Fly

Green bottle fly

House fly head

Hover fly

Blowfly maggots

Mosquito head

Midge

Horsefly head

Caddis fly

Flea jumping

Flies, Fleas and Mosquitoes

ANOPHELES MOSQUITO
Family *Culicidae*

COMMON FLEA
Family *Pulicidae*

HOUSE FLY
Family *Muscidae*

Mosquito head: long proboscis is inserted into prey to suck blood.

Flea head: backward facing cheek combs identify species.

Fly head: colorfully patterned compound eyes aid navigation.

MOSQUITO LIFE STAGES

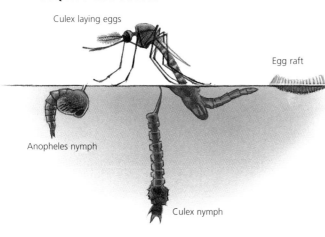

Culex laying eggs

Egg raft

Anopheles nymph

Culex nymph

CULEX MOSQUITO
Culex sp.

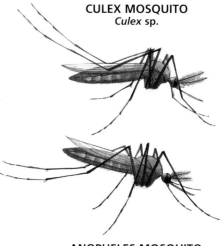

ANOPHELES MOSQUITO
Anopheles sp.

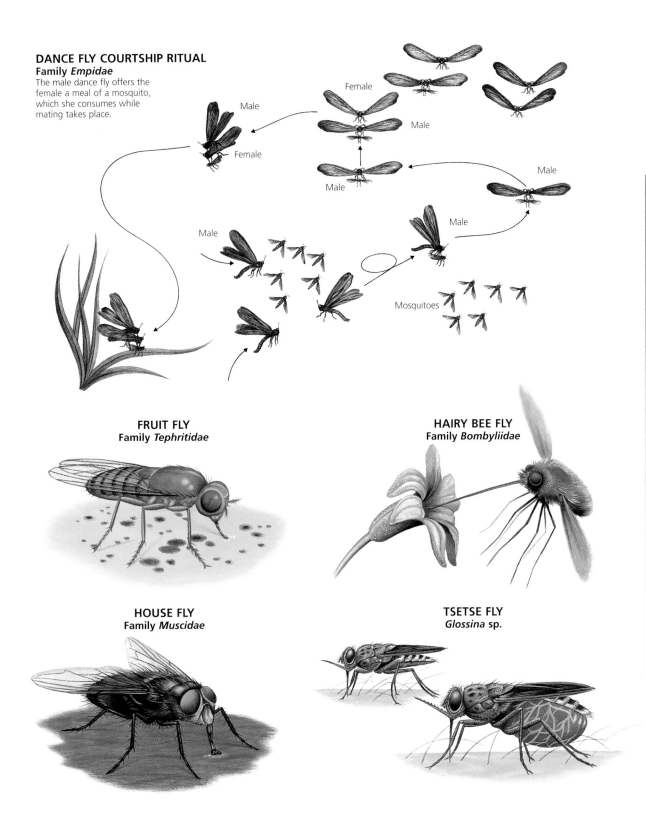

DANCE FLY COURTSHIP RITUAL
Family *Empidae*
The male dance fly offers the female a meal of a mosquito, which she consumes while mating takes place.

Female

Male

Female

Male

Male

Male

Male

Male

Male

Male

Female

Mosquitoes

FRUIT FLY
Family *Tephritidae*

HAIRY BEE FLY
Family *Bombyliidae*

HOUSE FLY
Family *Muscidae*

TSETSE FLY
Glossina sp.

Butterflies and Moths

BUTTERFLY CHRYSALIS (PUPA)

BUTTERFLY AND MOTH COMPARISON

Butterfly wings fold vertically when resting.

Moth wings are held horizontal when resting.

Thin, threadlike antennae with clubbed tips

Feathery or straight antennae without clubbed tips

Butterfly caterpillar

Moth caterpillar

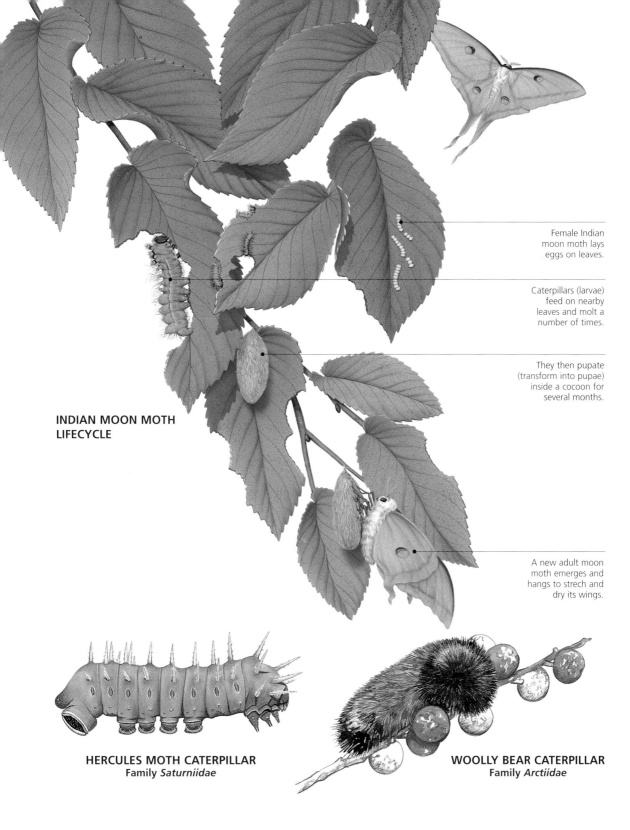

**INDIAN MOON MOTH
LIFECYCLE**

Female Indian moon moth lays eggs on leaves.

Caterpillars (larvae) feed on nearby leaves and molt a number of times.

They then pupate (transform into pupae) inside a cocoon for several months.

A new adult moon moth emerges and hangs to strech and dry its wings.

HERCULES MOTH CATERPILLAR
Family *Saturniidae*

WOOLLY BEAR CATERPILLAR
Family *Arctiidae*

Butterflies

HACKBERRY BUTTERFLY
Family *Nymphalidae*

SWALLOWTAIL
Family *Papilionidae*

PAINTED LADY BUTTERFLY
Family *Nymphalidae*

PIPEVINE
SWALLOWTAIL
Family *Papilionidae*

Anise swallowtail drinking nectar

Butterfly in flight

Butterfly at rest

Arthropods

INVERTEBRATES

DEADLY OR HARMLESS?

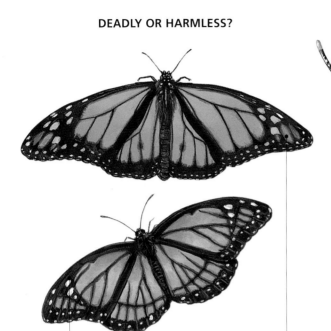

The harmless viceroy mimics the monarch so predators stay away.

The poisonous monarch butterfly has bold orange coloring as a warning.

88 BUTTERFLY
Diaethria sp.

MALAYAN LACEWING BUTTERFLY
Family *Nymphalidae*

BORDERED PATCH BUTTERFLY
Family *Nymphalidae*

WHITE ADMIRAL BUTTERFLY
Family *Nymphalidae*

Moths

INDIAN MOON MOTH
Family *Saturniidae*

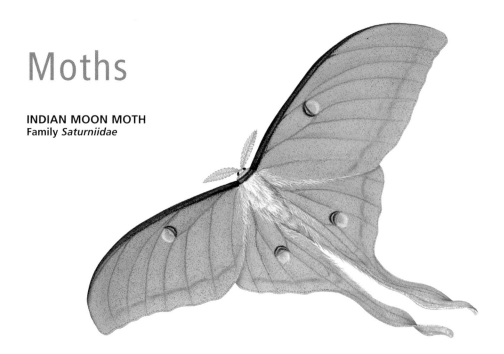

WHITE-LINED SPHINX MOTH
Family *Sphingidae*

HUMMINGBIRD MOTH
Family *Sphingidae*

Moth larva eating leaves

Mexican 'jumping beans' (moth larvae)

Sheep moth

FIVE-SPOTTED BURNET MOTH
Family *Zygaenidae*

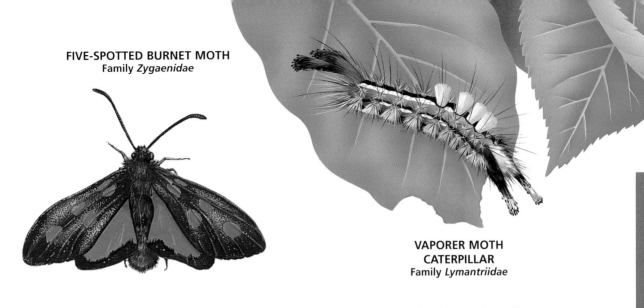

**VAPORER MOTH
CATERPILLAR**
Family *Lymantriidae*

CERISY'S SPHINX MOTH
Family *Sphingidae*

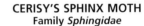

YELLOW EMPEROR MOTH
Family *Saturniidae*

**PUSS MOTH
CATERPILLAR**
Family *Notodontidae*

**MADAGASCAN
SUNSET MOTH**
Family *Uraniidae*

545

Bees, Wasps and Ants

Arthropods

INVERTEBRATES

SAND WASP
Family *Sphecidae*

BULLDOG ANT
Family *Formicidae*

BUMBLEBEE
Family *Apidae*

POTTER WASP
Family *Vespidae*

BUILDING A WASP NEST

Solitary queen begins nest.

She adds cells to lay eggs in.

First eggs hatch into workers, who continue to build cells.

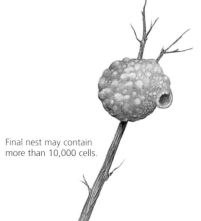

Final nest may contain more than 10,000 cells.

Honey bee head

Velvet ant (species of wasp)

Army ant

Hairy-legged mining bee

Digger wasp gathering nectar

Trap-jaw ant head

ANTS AND PLANTS

Bull's horn acacia
has broad,
hollow thorns.

Some species of
ants use them
as their home.

In return for this
hospitality, the ants
chase off possible
threats, such as
beetles and cows.

QUEENLY DUTIES

Only a queen ant is able to
lay eggs, which are taken
by a worker and tended
until adulthood.

Wasps and Ants

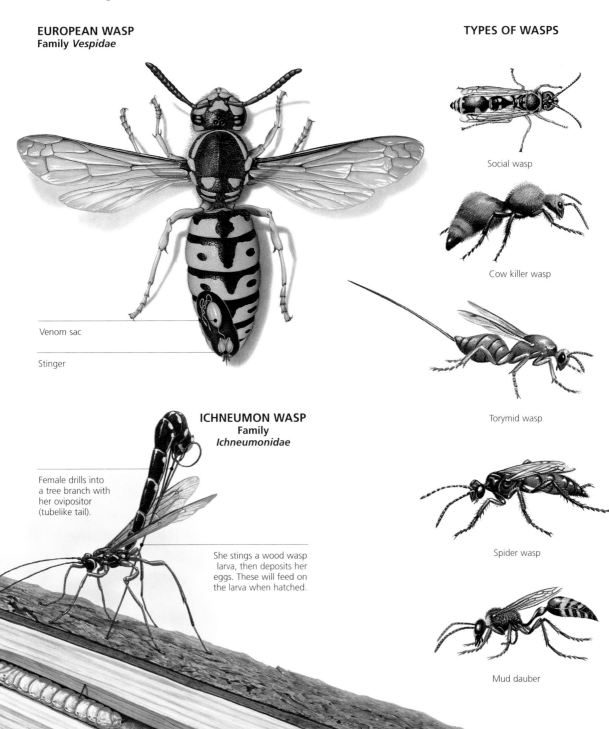

EUROPEAN WASP
Family *Vespidae*

Venom sac

Stinger

ICHNEUMON WASP
Family
Ichneumonidae

Female drills into
a tree branch with
her ovipositor
(tubelike tail).

She stings a wood wasp
larva, then deposits her
eggs. These will feed on
the larva when hatched.

TYPES OF WASPS

Social wasp

Cow killer wasp

Torymid wasp

Spider wasp

Mud dauber

TYPES OF BIG-HEADED ANTS

Worker

Soldier

Male

Queen

RED ANTS GREETING
Family *Formicidae*

AMAZON ANT
Family *Formicidae*

Pincers used for
fighting other ants

LEAF-CUTTER ANTS
Atta sp.

Inside the nest, the leaves
are chewed to a pulp, then
used as compost for the
special fungus the ants eat.

Small workers
remove any parasitic
flies from the leaves.

The leaf pieces are left
at the entrance to be
carried into the nest.

Medium-sized workers
cut pieces of leaves to
carry back to the nest.

Honey Bees

HONEY BEE DANCE

Worker bees returning from a food source dance to inform others in the hive where it is.

If the food source is close by, the dance is circular, the alignment showing the bee's direction relative to the sun.

If the source is further away, the dance is a figure-of-eight, the speed and number of 'waggles' at the center indicating distance.

GATHERING NECTAR

Bees gather nectar from flowers to provide food for the hive, storing it on their legs.

Providing nectar for bees is an effective way for plants to have pollen distributed to fertilize other flowers.

After converting the nectar into honey, bees store it in cells for the hive to share.

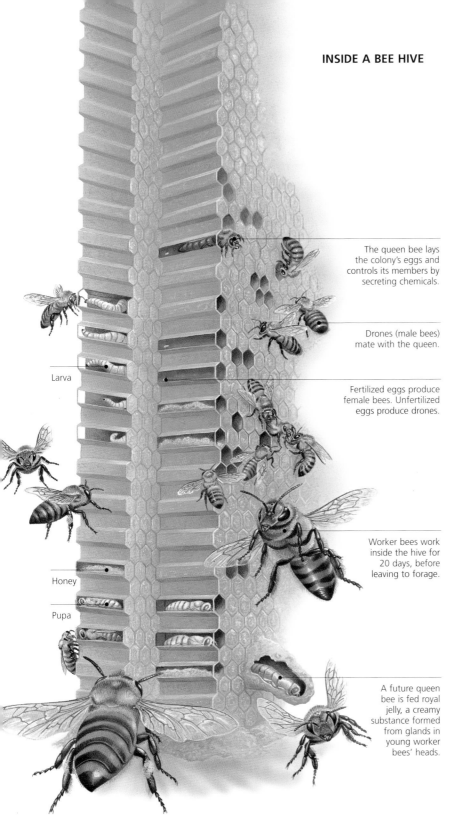

INSIDE A BEE HIVE

Larva

Honey

Pupa

The queen bee lays the colony's eggs and controls its members by secreting chemicals.

Drones (male bees) mate with the queen.

Fertilized eggs produce female bees. Unfertilized eggs produce drones.

Worker bees work inside the hive for 20 days, before leaving to forage.

A future queen bee is fed royal jelly, a creamy substance formed from glands in young worker bees' heads.

BEE EGG
Single egg laid in each cell

MOVING HIVES
Queen gathers a group of workers to leave the old hive.

The vulnerable queen is protected until a new home is found.

Scouts inform the others when a new home is found, like this beekeeper's box.

Classifying Animals

LEVELS OF CLASSIFICATION

Kingdom
There are five kingdoms, of which animals is one. Plants, fungi, protists and monerans are the others.

Phylum
First subdivision within the kingdom, linking broadly similar creatures. Vertebrates (chordata) form one phylum.

Class
Subdivision within the phylum linking similar orders. Mammals, birds, reptiles, amphibians and fishes are all classes.

Order
Subdivision within the class linking family groups by common characteristics. Marsupials and bats are orders.

Family
Subdivision within the order linking closely related individuals according to biology and behavior, like vipers or geckos.

Genus
Subdivision within the family, containing one or more species that are biologically very closely related, like salmons or tunas.

Species
Individuals within the genus that can interbreed and produce offspring that can interbreed, such as ostriches.

BROWN BEAR
Ursus arctos

FIRE SALAMANDER
Salamandra salamandra

BROWN BEAR	FIRE SALAMANDER
ANIMALIA (Kingdom)	**ANIMALIA** (Kingdom)
CHORDATA (Phylum)	**CHORDATA** (Phylum)
MAMMALIA (Class)	**AMPHIBIA** (Class)
CARNIVORA (Order)	**CAUDATA** (Order)
URSIDAE (Family)	**SALAMANDRIDAE** (Family)
URSUS (Genus)	**SALAMANDRA** (Genus)
ARCTOS (Species)	**SALAMANDRA** (Species)

Mammals

CLASS MAMMALIA MAMMALS

SUBCLASS PROTOTHERIA EGG-LAYING MAMMALS
ORDER MONOTREMATA MONOTREMES

Tachyglossidae	Echidnas
Ornithorhynchidae	Duck-billed platypus

SUBCLASS THERIA MODERN MAMMALS
INFRACLASS METATHERIA POUCHED MAMMALS
ORDER MARSUPIALIA MARSUPIALS

Didelphidae	American opossums

Microbiotheriidae	Colocolos
Caenolestidae	Shrew-opossums
Dasyuridae	Marsupial mice
Myrmecobiidae	Numbat
Thylacinidae	Thylacine
Notoryctidae	Marsupial mole
Peramelidae	Bandicoots
Peroryctidae	Spiny bandicoots
Vombatidae	Wombats
Phascolarctidae	Koala
Phalangeridae	Phalangers
Petauridae	Gliding phalangers
Pseudocheiridae	Ringtail possums
Burramyidae	Pygmy possums
Acrobatidae	Feathertails
Tarsipedidae	Honey possum
Macropodidae	Kangaroos, wallabies
Potoroidae	Bettongs

INFRACLASS EUTHERIA PLACENTAL MAMMALS
ORDER XENARTHRA ANTEATERS, SLOTHS AND ARMADILLOS

Myrmecophagidae	American anteaters
Bradypodidae	Three-toed sloths
Megalonychidae	Two-toed sloths
Dasypodidae	Armadillos

ORDER PHOLIDOTA PANGOLINS, SCALY ANTEATERS

Manidae	Pangolins, scaly anteater

ORDER INSECTIVORA INSECTIVORES

Solenodontidae	Solenodons
Tenrecidae	Tenrecs, otter shrews
Chrysochloridae	Golden moles
Erinaceidae	Hedgehogs, moonrats
Soricidae	Shrews
Talpidae	Moles, desmans

ORDER SCANDENTIA TREE SHREWS

Tupaiidae	Tree shrews

ORDER DERMOPTERA FLYING LEMURS (COLUGOS)

Cynocephalidae	Flying lemurs, colugos

ORDER CHIROPTERA BATS

Pteropodidae	Old World fruit bats
Rhinopomatidae	Mouse-tailed bats
Emballonuridae	Sheath-tailed bats
Craseonycteridae	Hog-nosed bat, bumblebee bat
Nycteridae	Slit-faced bats
Megadermatidae	False vampire bats
Rhinolophidae	Horseshoe bats
Hipposideridae	Old World leaf-nosed bats
Noctilionidae	Bulldog bats

Mormoopidae	Naked-backed bats
Phyllostomidae	New World leaf-nosed bats
Natalidae	Funnel-eared bats
Furipteridae	Smoky bats
Thyropteridae	Disk-winged bats
Myzopodidae	Old World sucker-footed bats
Vespertilionidae	Vespertilionid bats
Mystacinidae	New Zealand short-tailed bats
Molossidae	Free-tailed bats

ORDER PRIMATES — PRIMATES

Cheirogaleidae	Dwarf lemurs
Lemuridae	Large lemurs
Megaladapidae	Sportive lemurs
Indridae	Leaping lemurs
Daubentoniidae	Aye-aye
Loridae	Lorises, galagos
Tarsiidae	Tarsiers
Callitrichidae	Marmosets, tamarins
Cebidae	New World monkeys
Cercopithecidae	Old World monkeys
Hylobatidae	Gibbons
Hominidae	Great apes, humans

ORDER CARNIVORA — CARNIVORES

Canidae	Dogs, foxes
Ursidae	Bears, pandas
Procyonidae	Racoons and relatives
Mustelidae	Mustelids
Viverridae	Civets and relatives
Herpestidae	Mongooses
Hyaenidae	Hyenas
Felidae	Cats
Otariidae	Sealions
Odobenidae	Walrus
Phocidae	Seals

ORDER CETACEA — WHALES, DOLPHINS AND PORPOISES

Platanistidae	River dolphins
Delphinidae	Dolphins
Phocoenidae	Porpoises
Monodontidae	Narwhal, white whale
Physeteridae	Sperm whales
Ziphiidae	Beaked whales
Eschrichtiidae	Gray whale
Balaenopteridae	Rorquals
Balaenidae	Right whales

ORDER SIRENIA — SEA COWS

Dugongidae	Dugong
Trichechidae	Manatees

ORDER PROBOSCIDEA — ELEPHANTS

Elephantidae	Elephants

ORDER PERISSODACTYLA — ODD-TOED UNGULATES

Equidae	Horses
Tapiridae	Tapirs
Rhinocerotidae	Rhinoceroses

ORDER HYRACOIDEA — HYRAXES

Procaviidae	Hyraxes

ORDER TUBULIDENTATA — AARDVARK

Orycteropodidae	Aardvark

ORDER ARTIODACTYLA   EVEN-TOED UNGULATES

Suidae	Pigs
Tayassuidae	Peccaries
Hippopotamidae	Hippopotamuses
Camelidae	Camels, camelids
Tragulidae	Mouse deer
Moschidae	Musk deer
Cervidae	Deer

ORDER ARTIODACTYLA   EVEN-TOED UNGULATES

Giraffidae	Giraffe, okapi
Antilocapridae	Pronghorn
Bovidae	Cattle, antelopes, sheep and goats

ORDER RODENTIA   RODENTS

Aplodontidae	Mountain beaver
Sciuridae	Squirrels, marmots and relatives
Geomyidae	Pocket gophers
Heteromyidae	Pocket mice
Castoridae	Beavers
Anomaluridae	Scaly-tailed squirrels
Pedetidae	Spring hare
Muridae	Rats, mice, gerbils and relatives
Gliridae	Dormice
Seleviniidae	Desert dormouse
Zapodidae	Jumping mice
Dipodidae	Gerboas
Hystricidae	Old World porcupines
Erethizontidae	New World porcupines
Caviidae	Cavies and relatives
Hydrochaeridae	Capybara
Dinomyidae	Pacarana
Dasyproctidae	Agoutis, pacas
Chinchillidae	Chinchillas and relatives
Capromyidae	Hutias and relatives
Myocastoridae	Coypu

Octodontidae	Degus and relatives
Ctenomyidae	Tuco-tucos
Abrocomidae	Chinchilla-rats
Echymidae	Spiny rats
Thryonomyidae	Cane rats
Petromyidae	African rock-rat
Bathyergidae	African mole-rats
Ctenodactylidae	Gundis

ORDER LAGOMORPHA   LAGOMORPHS

Ochotonidae	Pikas
Leporidae	Rabbits, hares

ORDER MACROSCELIDEA   ELEPHANT SHREWS

Macroscelididae	Elephant shrews

Birds

CLASS AVES   BIRDS

ORDER STRUTHIONIFORMES   RATITES AND TINAMOUS

Struthionidae	Ostrich
Tinamidae	Tinamous
Rheidae	Rheas
Casuariidae	Cassowaries
Dromaiidae	Emu
Apterygidae	Kiwis

ORDER PROCELLARIIFORMES   ALBATROSSES AND PETRELS

Diomedeidae	Albatrosses
Procellariidae	Shearwaters
Hydrobatidae	Storm petrels
Pelecanoididae	Diving petrels

ORDER SPHENISCIFORMES   PENGUINS

Spheniscidae   Penguins

ORDER GAVIIFORMES   DIVERS

Gaviidae   Divers (loons)

ORDER PODICIPEDIFORMES   GREBES

Podicipedidae   Grebes

ORDER PELECANIFORMES   PELICANS AND RELATIVES

Phaethontidae	Tropicbirds
Pelecanidae	Pelicans
Phalacrocoracidae	Cormorants, anhingas
Sulidae	Gannets, boobies
Fregatidae	Frigatebirds

ORDER CICONIIFORMES   HERONS AND RELATIVES

Ardeidae	Herons, bitterns
Scopidae	Hammerhead
Ciconiidae	Storks
Balaenicipitidae	Whale-headed stork
Threskiornithidae	Ibises, spoonbills

ORDER PHOENICOPTERIFORMES   FLAMINGOS

Phoenicopteridae   Flamingos

ORDER FALCONIFORMES   RAPTORS

Accipitridae	Osprey, kites, hawks, eagles, Old World vultures, harriers, buzzards, harpies, buteonines
Sagittariidae	Secretarybird
Falconidae	Falcons, falconets, caracaras
Cathartidae	New World vultures

ORDER ANSERIFORMES   WATERFOWL AND SCREAMERS

Anatidae	Geese, swans, ducks
Anhimidae	Screamers

ORDER GALLIFORMES   GAMEBIRDS

Megapodiidae	Megapodes (mound-builders)
Cracidae	Chachalacas, guans, curassows
Phasianidae	Turkeys, grouse and relatives

ORDER OPISTHOCOMIFORMES   HOATZIN

Opisthocomidae   Hoatzin

ORDER GRUIFORMES · CRANES AND RELATIVES

Mesitornithidae	Mesites
Turnicidae	Hemipode-quails (button quails)
Pedionomidae	Collared hemipode
Gruidae	Cranes
Aramidae	Limpkins
Psophiidae	Trumpeters
Rallidae	Rails
Heliornithidae	Finfoots
Rhynochetidae	Kagus
Eurypygidae	Sunbittern
Cariamidae	Seriemas
Otididae	Bustards

ORDER CHARADRIIFORMES · WADERS AND SHOREBIRDS

Jacanidae	Jacanas
Rostratulidae	Painted snipe
Dromadidae	Crab plover
Haematopodidae	Oystercatchers
Ibidorhynchidae	Ibisbill
Recurvirostridae	Stilts, avocets
Burhinidae	Stone curlews (thick knees)
Glareolidae	Coursers, pratincoles
Charadriidae	Plovers, dotterels
Scolopacidae	Curlews, sandpipers, snipes
Thinocoridae	Seedsnipes
Chionididae	Sheathbills
Laridae	Gulls, terns, skimmers
Stercorariidae	Skuas, jaegers
Alcidae	Auks

ORDER COLUMBIFORMES · PIGEONS AND SANDGROUSE

Pteroclididae	Sandgrouse
Columbidae	Pigeons, doves

ORDER PSITTACIFORMES · PARROTS

Cacatuidae	Cockatoos
Psittacidae	Parrots

ORDER CUCULIFORMES · TURACOS AND CUCKOOS

Musophagidae	Turacos, louries (plantain-eaters)
Cuculidae	Cuckoos and relatives

ORDER STRIGIFORMES · OWLS

Tytonidae	Barn owls, bay owls
Strigidae	Hawk owls (true owls)

ORDER CAPRIMULGIFORMES · NIGHTJARS AND FROGMOUTHS

Steatornithidae	Oilbird
Podargidae	Frogmouths
Nyctibiidae	Potoos
Aegothelidae	Owlet nightjars
Caprimulgidae	Nightjars

ORDER APODIFORMES · SWIFTS AND HUMMINGBIRDS

Apodidae	Swifts
Hemiprocnidae	Crested swifts
Trochilidae	Hummingbirds

ORDER COLIIFORMES · MOUSEBIRDS

Coliidae	Mousebirds

ORDER TROGONIFORMES · TROGONS

Trogonidae	Trogons

ORDER CORACIIFORMES   KINGFISHERS AND RELATIVES

Alcedinidae	Kingfishers
Todidae	Todies
Momotidae	Motmots
Meropidae	Bee-eaters
Coraciidae	Rollers
Upupidae	Hoopoe
Phoeniculidae	Wood-hoopoes
Bucerotidae	Hornbills

ORDER PICIFORMES   WOODPECKERS AND BARBETS

Galbulidae	Jacamars
Bucconidae	Puffbirds
Capitonidae	Barbets
Ramphastidae	Toucans
Indicatoridae	Honeyguides
Picidae	Woodpeckers

ORDER PASSERIFORMES   PASSERINES OR PERCHING BIRDS

Suborder Eurylaimi   Broadbills and Pittas

Eurylaimidae	Broadbills
Philepittidae	Sunbirds, asitys
Pittidae	Pittas
Acanthisittidae	New Zealand wrens

Suborder Furnarii   Ovenbirds and relatives

Dendrocolaptidae	Woodcreepers
Furnariidae	Ovenbirds
Formicariidae	Antbirds
Rhinocryptidae	Tapaculos

Suborder Tyranni   Tyrant Flycatchers and Relatives

Tyrannidae	Tyrant flycatchers
Pipridae	Manakins
Cotingidae	Cotingas
Oxyruncidae	Sharpbills
Phytotomidae	Plantcutters

Suborder Oscines   Songbirds

Menuridae	Lyrebirds
Atrichornithidae	Scrub-birds
Alaudidae	Larks
Motacillidae	Wagtails, pipits
Hirundinidae	Swallows, martins
Campephagidae	Cuckoo-shrikes and relatives
Pycnonotidae	Bulbuls
Irenidae	Leafbirds, ioras, bluebirds
Laniidae	Shrikes
Vangidae	Vangas
Bombycillidae	Waxwings
Hypocoliidae	Hypocolius
Ptilogonatidae	Silky flycatchers
Dulidae	Palmchat
Prunellidae	Accentors, hedge-sparrows
Mimidae	Mockingbirds and relatives
Cinclidae	Dippers
Turdidae	Thrushes
Timaliidae	Babblers and relatives
Troglodytidae	Wrens
Sylviidae	Old World warblers
Muscicapidae	Old World flycatchers
Maluridae	Fairywrens and relatives
Acanthizidae	Australian warblers and relatives
Ephthianuridae	Australian chats
Orthonychidae	Logrunners and relatives
Rhipiduridae	Fantails
Monarchidae	Monarch flycatchers
Petroicidae	Australasian robins
Pachycephalidae	Whistlers and relatives
Aegithalidae	Long-tailed tits
Remizidae	Penduline tits

Paridae	True tits, chickadees, titmice
Sittidae	Nuthatches, sitellas, wallcreeper
Certhiidae	Holarctic treecreepers
Rhabdornithidae	Philippine treecreepers
Climacteridae	Australasian treecreepers
Dicaeidae	Flowerpeckers, pardalotes
Nectariniidae	Sunbirds
Zosteropidae	White-eyes
Meliphagidae	Honeyeaters
Vireonidae	Vireos
Emberizidae	Buntings, tanagers
Parulidae	New World wood warblers
Icteridae	Icterids (American blackbirds)
Fringillidae	Finches

Drepanididae	Hawaiian honeycreepers
Estrildidae	Estrildid finches
Ploceidae	Weavers
Passeridae	Old World sparrows
Sturnidae	Starlings, mynahs
Oriolidae	Orioles, figbirds
Dicruridae	Drongos
Callaeidae	New Zealand wattlebirds
Grallinidae	Magpie-larks
Artamidae	Wood swallows
Cracticidae	Bell magpies
Ptilonorhynchidae	Bowerbirds
Paradisaeidae	Birds-of-paradise
Corvidae	Crows, jays

Reptiles

CLASS REPTILIA   REPTILES

SUBCLASS EUREPTILIA
SUPERORDER TESTUDINE

ORDER TESTUDINATA	TURTLES, TERRAPINS AND TORTOISES

Suborder Pleurodira	Side-neck Turtles
Chelidae	Snake-neck turtles
Pelomedusidae	Helmeted side-neck turtles

Suborder Cryptodira	Hidden-necked Turtles

Superfamily Trionychoidea

Kinosternidae	Mud and musk turtles
Dermatemydidae	Mesoamerican river turtle
Carretochelyidae	Australian softshell turtle
Trionychidae	Holarctic and paleotropical softshell turtles

Superfamily Cheloniodea

Dermochelyidae	Leatherback sea turtles
Cheloniidae	Sea turtles

Superfamily Testudinoidea

Chelydridae	Snapping turtles
Emydidae	New World pond turtles and terrapins
Testudinidae	Tortoises
Bataguridae	Old World pond turtles

SUPERORDER ARCHOSAURIA

ORDER CROCODILIA	CROCODILIANS
Alligatoridae	Alligators and caimans
Crocodylidae	Crocodiles
Gavialidae	Gharials

SUPERORDER LEPIDOSAURIA

ORDER RHYNCHOCEPHALIA	TUATARA
Sphenodontidae	Tuatara

ORDER SQUAMATA — SQUAMATES

Suborder Amphisbaenia	Amphisbaenians

Bipedidae	Ajolotes
Amphisbaenidae	Worm lizards
Trogonophidae	Desert ringed lizards
Rhineuridae	Florida worm lizard

Suborder Iguania	Iguanid Lizards
Corytophanidae	Helmeted lizards
Crotaphytidae	Collared and leopard lizards
Hoplocercidae	Hoplocercids
Iguanidae	Iguanas
Opluridae	Madagascar iguanians
Phrynosomatidae	Scaly, sand and horned lizards
Polychrotidae	Anoloid lizards
Tropiduridae	Tropidurids
Agamidae	Agamid lizards
Chameleonidae	Chameleons

Suborder Scleroglossa	
Superfamily Gekkonoidea	

Eublepharidae	Eye-lash geckos
Gekkonidae	Geckos
Pygopodidae	Australasian flapfoots

Superfamily Scincoidea	
Xantusiidae	Night lizards
Lacertidae	Lacertids
Scincidae	Skinks
Dibamidae	Dibamids
Cordylidae	Girdle-tailed lizards
Gerrhosauridae	Plated lizards
Teiidae	Macroteiids
Gymnophthalmidae	Microteiids

Superfamily Anguoidea	
Xenosauridae	Knob-scaled lizards
Anguidae	Anguids; glass and alligator lizards
Helodermatidae	Beaded lizards
Varanidae	Monitor lizards
Lanthanotidae	Earless monitor lizard

Suborder Serpentes	Snakes
Infraorder Scolecophidia	
Anomalepididae	Blind wormsnakes
Typhlopidae	Blind snakes
Leptotyphlopidae	Thread snakes

Infraorder Alethinophidia	
Anomochelidae	Stump heads
Aniliidae	Coral pipesnakes
Cylindrophidae	Asian pipesnakes
Uropeltidae	Shield tails
Xenopeltidae	Sunbeam snake
Loxocemidae	Dwarf boa
Boidae	Pythons and boas

Ungaliophiidae	Ungaliophiids
Bolyeriidae	Round Island snakes
Tropidophiidae	Woodsnakes
Acrochordidae	File snakes
Atractaspididae	Mole vipers
Colubridae	Harmless and rear-ranged snakes
Elapidae	Cobras, kraits, coral snakes and sea snakes
Viperidae	Adders and vipers

Amphibians

CLASS AMPHIBIA   AMPHIBIANS

SUBCLASS LISSAMPHIBIA

ORDER CAUDATA	SALAMANDERS AND NEWTS

Suborder Sirenoidea

Sirenidae	Sirens

Suborder Cryptobranchoidea

Cryptobranchidae	Hellbenders and giant salamanders
Hynobiidae	Hynobiids

Suborder Salamandroidea

Amphiumidae	Amphiumas (congo eels)
Plethodontidae	Lungless salamanders
Rhyacotritonidae	Torrent salamanders
Proteidae	Mudpuppies, waterdogs, and the olm
Salamandridae	Salamandrids

Ambystomatidae	Mole salamanders
Dicamptodontidae	Dicamptodontids

ORDER GYMNOPHIONA   CAECILIANS

Rhinatrematidae	South American tailed caecilians
Ichthyophiidae	Ichthyophiids
Uraeotyphlidae	Uraeotyphlids
Scolecomorphidae	Scolecomorphids
Caeciliidae	Caeciliids and aquatic caecilians

ORDER ANURA   FROGS AND TOADS

Ascaphidae	"Tailed" frogs
Leiopelmatidae	New Zealand frogs
Bombinatoridae	Fire-bellied toads and allies

Discoglossidae	Discoglossid frogs
Pipidae	Pipas and "clawed" frogs
Rhinophrynidae	Cone-nosed frog
Megophryidae	Megophryids
Pelodytidae	Parsley frogs
Pelobatidae	Spadefoots

Superfamily Bufonoidea

Allophrynidae	Allophrynid frog
Brachycephalidae	Saddleback frogs
Bufonidae	Toads
Helophrynidae	Ghost frogs
Leptodactylidae	Neotropical frogs
Myobatrachidae	Australasian frogs
Sooglossiidae	Seychelles frogs
Rhinodermatidae	Darwin's frogs
Hylidae	Hylid treefrogs
Pelodryadidae	Australasian treefrogs
Centrolenidae	Glass frogs
Pseudidae	Natator frogs
Dendrobatidae	Dart-poison frogs

Superfamily Ranoidea

Microhylidae	Microhylids
Hemisotidae	Shovel-nosed frogs
Arthroleptidae	Squeakers
Ranidae	Ranid frogs
Hyperoliidae	Reed and lily frogs
Rhacophoridae	Rhacaphorid treefrogs

Fishes

SUPERCLASS AGNATHA — JAWLESS FISHES

CLASS MYXINI — HAGFISHES

ORDER MYXINIFORMES	HAGFISHES
Myxinidae	Hagfishes

CLASS CEPHALASPIDOMORPHI LAMPREYS AND RELATIVES

ORDER PETROMYZONTIFORMES	LAMPREYS
Petromyzontidae	Northern lampreys
Geotriidae	Pouched lampreys
Mordaciidae	Shorthead lampreys

SUPERCLASS GNATHOSTOMATA — JAWED FISHES

CLASS CHONDRICHTHYES CARTILAGINOUS FISHES

SUBCLASS ELASMOBRANCHII — SHARKS AND RAYS

ORDER HEXANCHIFORMES	SIX- AND SEVENGILL SHARKS
Chlamydoselachidae	Frill shark
Hexanchidae	Sixgill, sevengill (cow) sharks

ORDER SQUALIFORMES	DOGFISH SHARKS AND RELATIVES
Echinorhinidae	Bramble sharks
Squalidae	Dogfish sharks
Oxynotidae	Roughsharks

ORDER PRISTIOPHORIFORMES	SAWSHARKS
Pristiophoridae	Sawsharks

ORDER HETERODONTIFORMES	HORN (BULLHEAD) SHARKS
Heterodontidae	Horn (bullhead) sharks

ORDER ORECTOLOBIFORMES	CARPETSHARKS AND RELATIVES
Parascylliidae	Collared carpetsharks
Brachaeluridae	Blind sharks
Orectolobidae	Wobbegongs
Hemiscylliidae	Longtail carpetsharks
Stegostomatidae	Zebra shark
Ginglymostomatidae	Nurse sharks
Rhincodontidae	Whale shark

ORDER LAMNIFORMES	MACKEREL SHARKS AND RELATIVES
Odontaspididae	Sand tigers (gray nurse sharks)
Mitsukurinidae	Goblin shark
Pseudocarchariidae	Crocodile shark
Megachasmidae	Megamouth shark
Alopiidae	Thresher sharks
Cetorhinidae	Basking shark
Lamnidae	Mackerel sharks (makos)

ORDER CARCHARHINIFORMES	GROUND SHARKS
Scyliorhinidae	Catsharks
Proscylliidae	Finback catsharks
Pseudotriakidae	False catshark
Leptochariidae	Barbled houndshark
Triakidae	Houndsharks
Hemigaleidae	Weasel sharks
Carcharhinidae	Requiem (whaler) sharks
Sphyrnidae	Hammerhead sharks

ORDER SQUATINIFORMES ANGEL SHARKS

Squatinidae	Angel sharks

ORDER RHINOBATIFORMES   SHOVELNOSE RAYS

Platyrhinidae	Platyrhinids
Rhinobatidae	Guitarfishes
Rhynchobatidae	Sharkfin guitarfishes

ORDER RAJIFORMES   SKATES

Rajidae	Skates

ORDER PRISTIFORMES   SAWFISHES

Pristidae	Sawfishes

ORDER TORPEDINIFORMES   ELECTRIC RAYS

Torpedinidae	Electric rays
Hypnidae	Coffinrays
Narcinidae	Narcinids
Narkidae	Narkids

ORDER MYLIOBATIFORMES   STINGRAYS AND RELATIVES

Potamotrygonidae	River rays
Dasyatididae	Stingrays
Urolophidae	Round stingrays
Gymnuridae	Butterfly rays
Hexatrygonidae	Sixgill rays
Myliobatididae	Eagle rays
Rhinopteridae	Cownose rays
Mobulidae	Mantas, devil rays

SUBCLASS HOLOCEPHALI   CHIMAERAS
ORDER CHIMAERIFORMES   CHIMAERAS

Callorhynchidae	Plownose chimaeras (Elephant fishes)
Chimaeridae	Shortnose chimaeras (Ghost sharks, ratfishes)
Rhinochimaeridae	Longnose chimaeras (spookfishes)

CLASS OSTEICHTHYES   BONY FISHES

SUBCLASS SARCOPTERYGII   LUNGFISHES AND COELACANTH (FLESHY-FINNED FISHES)

ORDER CERATODONTIFORMES   LUNGFISHES

Ceratodontidae	Australian lungfish

ORDER LEPIDOSIRENIFORMES   LUNGFISHES

Lepidosirenidae	South American lungfish
Protopteridae	African lungfishes

ORDER COELACANTHIFORMES   COELACANTH

Latimeriidae	Coelacanth (gombessa)

SUBCLASS ACTINOPTERYGII   RAY-FINNED FISHES
ORDER POLYPTERIFORMES BICHIRS

Polypteridae	Bichirs

ORDER ACIPENSERIFORMES   STURGEONS AND RELATIVES

Acipenseridae	Sturgeons
Polyodontidae	Paddlefishes

ORDER LEPISOSTEIFORMES GARS

Lepisosteidae	Gars

ORDER AMIIFORMES   BOWFIN

Amiidae	Bowfin

GROUP TELEOSTEI   TELEOSTS

ORDER HIODONTIFORMES   MOONEYES

Hiodontidae   Mooneyes

ORDER OSTEOGLOSSIFORMES   BONYTONGUES AND RELATIVES

Osteoglossidae	Bonytongues
Pantodontidae	Freshwater butterflyfish
Notopteridae	Featherbacks
Mormyridae	Elephantfishes
Gymnarchidae	Aba

ORDER ELOPIFORMES   TARPONS AND RELATIVES

Elopidae	Tenpounders (ladyfishes)
Megalopidae	Tarpons

ORDER ALBULIFORMES   BONEFISHES AND RELATIVES

Albulidae   Bonefishes

ORDER NOTACANTHIFORMES   SPINY EELS AND RELATIVES

Halosauridae	Halosaurs
Notacanthidae	Spiny eels

ORDER ANGUILLIFORMES   EELS

Anguillidae	Freshwater eels
Heterenchelyidae	Shortfaced eels
Moringuidae	Spaghetti eels
Chlopsidae	False moray eels
Myrocongridae	Myrocongrid
Muraenidae	Moray eels
Nemichthyidae	Snipe eels
Muraenesocidae	Pike eels
Synaphobranchidae	Cutthroat eels
Ophichthidae	Snake eels, worm eels
Nettastomatidae	Duckbill eels
Colocongridae	Colocongrids
Congridae	Conger eels, garden eels
Derichthyidae	Narrowneck eels, spoonbill eels
Serrivomeridae	Sawtooth eels
Cyematidae	Bobtail snipe eels

Saccopharyngidae	Swallowers
Eurypharyngidae	Gulper eels
Monognathidae	Singlejaw eels

ORDER CLUPEIFORMES   SARDINES, HERRINGS, AND ANCHOVIES

Denticipitidae	Denticle herring
Clupeidae	Sardines, herrings, pilchards
Engraulididae	Anchovies
Chirocentridae	Wolf herrings

ORDER GONORYNCHIFORMES   MILKFISH, BEAKED SALMONS AND RELATIVES

Chanidae	Milkfish
Gonorynchidae	Beaked salmons
Kneriidae	Shellears
Phractolaemidae	Hingemouth

ORDER CYPRINIFORMES   CARPS, MINNOWS AND RELATIVES

Cyprinidae	Carps, minnows
Psilorhynchidae	Psilorhynchids
Balitoridae	Hillstream loaches
Cobitidae	Loaches
Gyrinocheilidae	Algae eaters
Catostomidae	Suckers

ORDER CHARACIFORMES   CHARACINS AND RELATIVES

Citharinidae	Citharinids
Distichodontidae	Distichodontins, African fin eaters
Hepsetidae	African pike characin
Erythrinidae	Trahiras and relatives
Ctenoluciidae	American pike characins
Lebiasinidae	Voladoras, pencil fishes
Characidae	African tetras (characins), characins, tetras, brycons, piranhas and relatives
Curimatidae	Toothless characins
Prochilodontidae	Flannelmouth characins
Anostomidae	Anostomins, leporinins
Hemiodontidae	Hemiodontids
Chilodontidae	Headstanders
Gasteropelecidae	Freshwater hatchetfishes

Classifying Animals

FACT FILE

ORDER SILURIFORMES — CATFISHES AND KNIFEFISHES

Suborder Siluroidei

Diplomystidae	Velvet catfishes
Nematogenyidae	Mountain catfish
Trichomycteridae	Spinyhead catfishes, candirus
Callichthyidae	Plated catfishes
Scoloplacidae	Spinynose catfishes
Astroblepidae	Andes catfishes
Loricariidae	Armored suckermouth catfishes
Ictaluridae	Bullhead catfishes, madtoms
Bagridae	Bagrid catfishes
Cranoglanididae	Armorhead catfishes
Siluridae	Sheatfishes, wels, glass catfishes
Schilbidae	Schilbid catfishes
Pangasiidae	Pangasiid catfishes
Amblycipitidae	Torrent catfishes
Amphiliidae	Loach catfishes
Akysidae	Asian banjo catfishes
Sisoridae	Hillstream catfishes
Clariidae	Labyrinth (airbeathing) catfishes
Heteropneustidae	Airsac catfishes
Chacidae	Square head (angler) catfishes
Olyridae	Olyrid catfishes
Malapteruridae	Electric catfishes
Ariidae	Sea catfishes
Plotosidae	Eeltail catfishes
Mochokidae	Upsidedown catfishes (squeakers)
Doradidae	Thorny catfishes
Auchenipteridae	Wood catfishes
Pimelodidae	Longwhisker, shovelnose catfishes
Ageneiosidae	Bottlenose (barbelless) catfishes
Helogenidae	Helogene catfishes
Cetopsidae	Carneros, shark (whale) catfishes
Hypophthalmidae	Loweye catfishes
Aspredinidae	Banjo catfishes

Suborder Gymnotoidei

Sternopygidae	Longtail knifefishes
Rhamphichthyidae	Longsnout knifefishes
Hypopomidae	Bluntnose knifefishes
Apternotidae	Black knifefishes
Gymnotidae	Banded knifefishes
Electrophoridae	Electric eel (electric knifefish)

ORDER ARGENTINIFORMES — HERRINGS, SMELTS, BARRELEYES AND RELATIVES

Argentinidae	Herring smelts
Microstomatidae	Deepsea smelts
Opisthoproctidae	Barreleyes (spookfishes)
Alepocephalidae	Slickheads
Bathylaconidae	Bathylaconids
Platytroctidae	Tubeshoulders

ORDER SALMONIFORMES — SALMONS, SMELTS AND RELATIVES

Osmeridae	Northern smelts, ayu
Salangidae	Icefishes (noodlefishes)
Retropinnidae	Southern smelts
Galaxiidae	Galaxiids
Lepidogalaxiidae	Salamanderfish
Salmonidae	Trouts, salmons, chars, whitefishes

ORDER ESOCIFORMES — PIKES, PICKERELS AND RELATIVES

Esocidae	Pikes
Umbridae	Mudminnows

ORDER ATELEOPODIFORMES — JELLYNOSE FISHES

Ateleopodidae	Jellynose fishes

ORDER STOMIIFORMES — DRAGONFISHES, LIGHTFISHES AND RELATIVES

Gonostomatidae	Lightfishes, bristlemouths
Sternoptychidae	Hatchetfishes and relatives
Phosichthyidae	Lightfishes
Stomiidae	Viperfishes, dragonfishes, snaggletooths, loosejaws

ORDER AULOPIFORMES	LIZARDFISHES AND RELATIVES
Aulopidae	Aulopids
Bathysauridae	Bathysaurids
Chlorophthalmidae	Greeneyes
Ipnopidae	Ipnopids, tripod fishes
Synodontidae	Lizardfishes, Bombay duck
Scopelarchidae	Pearleyes
Notosudidae	Waryfishes
Giganturidae	Telescope fishes
Paralepididae	Barracudinas
Anotopteridae	Daggertooth
Evermannellidae	Sabertooth fishes
Omosudidae	Omosudid
Alepisauridae	Lancetfishes
Pseudotrichonotidae	Pseudotrichonotids

ORDER MYCTOPHIFORMES	LANTERNFISHES AND RELATIVES
Neoscopelidae	Neoscopelids
Myctophidae	Lanternfishes

ORDER LAMPRIDIFORMES	OARFISHES AND RELATIVES
Lampridae	Opahs
Veliferidae	Velifers
Lophotidae	Crestfishes
Radiicephalidae	Inkfishes
Trachipteridae	Ribbonfishes
Regalecidae	Oarfishes
Stylephoridae	Tube eye

ORDER POLYMIXIIFORMES	BEARDFISHES
Polymixiidae	Beardfishes

ORDER PERCOPSIFORMES	TROUTPERCHES AND RELATIVES
Percopsidae	Troutperches
Aphredoderidae	Pirate perch
Amblyopsidae	Cavefishes

ORDER GADIFORMES	CODS, HAKES AND RELATIVES
Muraenolepididae	Muraenolepidids
Moridae	Morid cods (moras)
Melanonidae	Melanonids
Euclichthyidae	Euclichthyid
Bregmacerotidae	Codlets
Gadidae	Codfishes, haddocks and relatives
Merlucciidae	Hakes
Steindachneriidae	Steindachneriid
Macrouridae	Rattails (grenadiers)

ORDER OPHIDIIFORMES	CUSKEELS, PEARLFISHES AND RELATIVES
Ophidiidae	Cuskeels, brotulas and relatives
Carapidae	Pearlfishes
Bythitidae	Livebearing brotulas
Aphyonidae	Aphyonids

ORDER BATRACHOIDIFORMES	TOADFISHES AND MIDSHIPMEN
Batrachoididae	Toadfishes and midshipmen

ORDER LOPHIIFORMES	ANGLERFISHES, GOOSEFISHES AND FROGFISHES
Lophiidae	Goosefishes (monkfishes)
Antennariidae	Frogfishes (shallow anglerfishes)
Tetrabrachiidae	Humpback angler
Lophichthyidae	Lophichthyid
Brachionichthyidae	Warty anglers (hand fishes)
Chaunacidae	Seatoads (gapers, coffinfishes)
Ogcocephalidae	Batfishes
Caulophrynidae	Fanfin anglers
Ceratiidae	Seadevils
Gigantactinidae	Whipnose anglers
Neoceratiidae	Needlebeard angler
Linophrynidae	Netdevils
Oneirodidae	Dreamers
Thaumatichthyidae	Wolftrap angler
Centrophrynidae	Hollowchin anglers
Diceratiidae	Double anglers
Himantolophidae	Footballfishes
Melanocetidae	Blackdevils

ORDER GOBIESOCIFORMES — CLINGFISHES AND RELATIVES

Gobiesocidae — Clingfishes
Callionymidae — Dragonets
Draconettidae — Draconettids

ORDER CYPRINODONTIFORMES — KILLIFISHES AND RELATIVES

Rivulidae — South American annuals
Aplocheilidae — African annuals
Profundulidae — Profundulids
Fundulidae — Killifishes
Valenciidae — Valenciids
Anablepidae — Foureyed fishes (cuatro ojos)
Poeciliidae — Livebearers (guppies, etc.)
Goodeidae — Goodeids
Cyprinodontidae — Pupfishes

ORDER BELONIFORMES — RICEFISHES, FLYINGFISHES AND RELATIVES

Adrianichthyidae — Ricefishes
Exocoetidae — Flyingfishes
Hemiramphidae — Halfbeaks
Belonidae — Needlefishes
Scomberesocidae — Sauries

ORDER ATHERINIFORMES — SILVERSIDES, RAINBOWFISHES AND RELATIVES

Atherinidae — Silversides, topsmelts, grunions
Dentatherinidae — Dentatherinid
Notocheiridae — Surf silversides
Melanotaeniidae — Rainbowfishes
Pseudomugilidae — Blue eyes
Telmatherinidae — Sailfin silversides
Phallostethidae — Priapium fishes

ORDER STEPHANOBERYCIFORMES — PRICKLEFISHES AND RELATIVES

Stephanoberycidae — Pricklefishes
Melamphaidae — Bigscale fishes (ridgeheads)
Gibberichthyidae — Gibberfishes
Hispidoberycidae — Bristlyskin
Rondeletiidae — Orangemouth whalefishes
Barbourisiidae — Redvelvet whalefish
Cetomimidae — Flabby whalefishes
Mirapinnidae — Hairyfish, tapetails
Megalomycteridae — Mosaicscale (bignose) fishes

ORDER BERYCIFORMES — SQUIRRELFISHES AND RELATIVES

Holocentridae — Squirrelfishes (soldier fishes)
Berycidae — Alfoncinos
Anoplogasteridae — Fangtooth fishes
Monocentrididae — Pineapple (pinecone) fishes
Anomalopidae — Flashlight (lanterneye) fishes
Trachichthyidae — Roughies (slimeheads)
Diretmidae — Spinyfins

ORDER ZEIFORMES — DORIES AND RELATIVES

Parazenidae — Parazen
Macrurocyttidae — Macrurocyttids
Zeidae — Dories
Oreosomatidae — Oreos
Grammicolepididae — Tinselfishes
Caproidae — Boarfishes

ORDER GASTEROSTEIFORMES — STICKLEBACKS, PIPEFISHES AND RELATIVES

Hypoptychidae — Sand eel
Aulorhynchidae — Tubesnouts
Gasterosteidae — Sticklebacks
Indostomidae — Paradox fish
Pegasidae — Seamoths
Aulostomidae — Trumpetfishes
Fistulariidae — Cornetfishes
Macroramphosidae — Snipefishes
Centriscidae — Shrimpfishes (razorfishes)
Solenostomidae — Ghost pipefishes
Syngnathidae — Pipefishes, seahorses

ORDER SYNBRANCHIFORMES — SWAMPEELS AND RELATIVES

Synbranchidae	Swampeels
Mastacembelidae	Spiny eels
Chaudhuriidae	Chaudhuriids

ORDER SCORPAENIFORMES — SCORPIONFISHES AND RELATIVES

Scorpaenidae	Scorpionfishes (rockfishes), stonefishes

Caracanthidae	Coral crouchers
Aploactinidae	Velvetfishes
Pataecidae	Prowfishes
Gnathanacanthidae	Red velvetfish
Congiopodidae	Pigfishes (horsefishes)
Triglidae	Sea robins (gurnards)
Dactylopteridae	Helmet (flying) gurnards

Platycephalidae	Flatheads
Bembridae	Deepwater flatheads
Hoplichthyidae	Spiny (ghost) flatheads
Anoplopomatidae	Sablefish, skilfish
Hexagrammidae	Greenlings
Zaniolepididae	Combfishes
Normanichthyidae	Southern sculpin
Rhamphocottidae	Grunt sculpin
Ereuniidae	Deepwater sculpins
Cottidae	Sculpins
Comephoridae	Baikal oilfishes
Abyssocottidae	Abyssocottids
Hemitripteridae	Hemitripterids
Psychrolutidae	Flatheads (blob fishes)
Bathylutichthyidae	Bathylutichthyids
Agonidae	Poachers
Cyclopteridae	Lumpfishes
Liparidae	Snailfishes, lumpsuckers

ORDER PERCIFORMES — PERCHES AND RELATIVES

Suborder Percoidei

Ambassidae	Glassfishes
Acropomatidae	Acropomatids
Epigonidae	Epigonids
Scrombropidae	Gnomefishes
Symphysanodontidae	Symphsanodontids
Caesioscorpididae	Caesioscorpidids
Polyprionidae	Wreckfishes
Dinopercidae	Cavebasses
Centropomidae	Snooks
Latidae	Giant perches
Percichthyidae	South temperate basses
Moronidae	North temperate basses
Serranidae	Sea basses, groupers
Callanthiidae	Rosey perches
Pseudochromidae	Dottybacks
Grammatidae	Basslets
Opistognathidae	Jawfishes
Plesiopidae	Roundheads
Notograptidae	Bearded snakeblennies
Pholidichthyidae	Convict blennies
Cepolidae	Bandfishes
Glaucosomatidae	Pearl perches
Terapontidae	Grunters
Banjosidae	Banjosids
Kuhliidae	Aholeholes
Centrarchidae	Sunfishes
Percidae	Perches, darters
Priacanthidae	Bigeyes (catalufas)
Apogonidae	Cardinalfishes
Dinolestidae	Dinolestid
Sillaginidae	Smelt whitings (whitings)
Malacanthidae	Tilefishes
Labracoglossidae	Labracoglossids
Lactariidae	False trevrelatives
Pomatomidae	Bluefish (tailor)
Menidae	Moonfish
Leiognathidae	Ponyfishes (slipmouths)
Bramidae	Pomfrets
Caristiidae	Manefishes
Arripidae	Australian salmons
Emmelichthyidae	Rovers
Lutjanidae	Snappers (emperors)
Lobotidae	Tripletails

Gerreidae	Mojarras (silver biddies)
Haemulidae	Grunts
Inermiidae	Bonnetmouths
Sparidae	Porgies (breams)
Centracanthidae	Centracanthids
Lethrinidae	Scavengers (emperors)
Nemipteridae	Threadfin, monocle breams
Sciaenidae	Drums, croakers
Mullidae	Goatfishes
Monodactylidae	Moonfishes (fingerfishes)
Pempherididae	Sweepers
Leptobramidae	Beachsalmon
Bathyclupeidae	Bathyclupeids
Toxotidae	Archerfishes
Coracinidae	Galjoenfishes
Kyphosidae	Seachubs
Girellidae	Nibblers (blackfishes)
Scorpididae	Halfmoons, mado (sweeps)
Microcanthidae	Stripey
Ephippidae	Spadefishes
Scatophagidae	Scats
Chaetodontidae	Butterflyfishes
Pomacanthidae	Angelfishes
Enoplosidae	Oldwife
Pentacerotidae	Armorheads
Oplegnathidae	Knifejaws
Icosteidae	Ragfish
Kurtidae	Nurseryfishes
Scombrolabracidae	Scombrolabracid

Suborder Carangoidei

Nematistiidae	Roosterfish
Carangidae	Jacks (trevallies)
Echeneidae	Remoras
Rachycentridae	Cobia (black kingfish)
Coryphaenidae	Dolphins

Suborder Cirrhitoidei

Cirrhitidae	Hawkfishes
Chironemidae	Kelpfishes
Aplodactylidae	Aplodactylids
Cheilodactylidae	Morwongs
Latrididae	Trumpeters

Suborder Mugiloidei

Mugilidae	Mullets

Suborder Polynemoidei

Polynemidae	Threadfins

Suborder Labroidei

Chichlidae	Cichlids
Embiotocidae	Surfperches
Pomacentridae	Damselfishes, anemonefishes
Labridae	Wrasses
Odacidae	Rock whitings
Scaridae	Parrotfishes

Suborder Zoarcoidei

Bathymasteridae	Ronquils
Zoarcidae	Eelpouts
Stichaeidae	Pricklebacks
Cryptacanthodidae	Wrymouths
Pholididae	Gunnels
Anarhichadidae	Wolffishes
Ptilichthyidae	Quillfishes
Zaproridae	Prowfish
Scytalinidae	Graveldiver

Suborder Notothenioidei

Bovichtidae	Thornfishes
Nototheniidae	Nototheniids
Harpagiferidae	Plunderfishes
Bathydraconidae	Antarctic dragonfishes
Channichthyidae	Icefishes

Suborder Trachinoidei

Chiasmodontidae	Swallowers
Champsodontidae	Champsodontids
Trichodontidae	Sandfishes
Trachinidae	Weeverfishes
Uranoscopidae	Stargazers
Trichonotidae	Sanddivers
Creediidae	Sandburrowers
Leptoscopidae	Leptoscopids
Percophidae	Duckbills
Pinguipedidae	Sandperches
Cheimarrhichthyidae	Torrentfish
Ammodytidae	Sandlances

Suborder Blennioidei

Tripterygiidae	Triplefins
Dactyloscopidae	Sand stargazers
Labrisomidae	Labrisomids
Clinidae	Kelp blennies (kelpfishes)
Chaenopsidae	Tube blennies
Blenniidae	Combtooth blennies

Suborder Gobiodei

Rhyacichthyidae	Loach goby
Odontobutidae	Freshwater Asian gobies
Gobiidae	Gobies, sleepers (gudgeons)

Suborder Acanthuroidei

Siganidae	Rabbitfishes
Luvaridae	Louvar
Zanclidae	Moorish idol
Acanthuridae	Surgeonfishes (tangs)

Suborder Scombroidei

Sphyraenidae	Barracudas
Gempylidae	Snake mackerels, gemfishes
Trichiuridae	Cutlassfishes
Scombridae	Mackerels, tunas, bonitos
Xiphiidae	Swordfish
Istiophoridae	Billfishes

Suborder Stromateoidei

Amarsipidae	Amarsipids
Centrolophidae	Medusafishes
Nomeidae	Driftfishes
Ariommatidae	Ariommatids
Tetragonuridae	Squaretails
Stromateidae	Butterfishes

Suborder Anabantoidei

Badidae	Chameleonfish
Nandidae	Leaffishes
Pristolepidae	False leaffishes
Channidae	Snakeheads
Anabantidae	Climbing gouramies
Belontiidae	Gouramies
Helostomatidae	Kissing gouramie
Osphronemidae	Giant gouramie
Luciocephalidae	Pikehead

ORDER PLEURONECTIFORMES   FLATFISHES

Psettodidae	Spiny flatfishes
Citharidae	Citharids
Bothidae	Lefteye flounders
Pleuronectidae	Tonguefishes (tongue soles)
Soleidae	Soles
Achiridae	American soles

ORDER TETRAODONTIFORMES   TRIGGERFISHES AND RELATIVES

Triacanthodidae	Spikefishes
Triacanthidae	Triplespines
Balistidae	Triggerfishes
Monacanthidae	Filefishes (leatherjackets)
Aracanidae	Robust boxfishes
Ostraciidae	Boxfishes, cowfishes, trunkfishes
Triodontidae	Pursefish (three-toothed puffer)
Tetraodontidae	Pufferfishes (toados)
Diodontidae	Porcupinefishes
Molidae	Ocean sunfishes (molas)

Invertebrates

PHYLUM PORIFERA — SPONGES

Class Calcarea	Calcareous sponges
Class Hexactinellida	Glass sponges
Class Demospongiae	Demosponges

PHYLUM CHORDATA — INVERTEBRATE CHORDATES

Class Urochordata	Sea squirts or tunicates
Class Cephalochardata	Lancelets

PHYLUM CNIDARIA — HYDRAS, JELLYFISH, SEA ANEMONES, CORALS

Class Hydrozoa	Hydras and freshwater jellyfish
Class Scyphozoa	Jellyfish
Class Cubozoa	Jellyfish
Class Anthozoa	Sea anemones, corals, sea fans, sea pansies

PHYLUM CTENOPHORA — SEA WALNUTS OR COMB JELLIES

Class Tentaculata
Class Nuda

PHYLUM PLATYHELMINTHES — FLATWORMS

Class Trematoda	Flukes
Class Monogenea	
Class Cestoda	Tapeworms
Class Turbellaria	Free-living flatworms
Subclass Archoophora	Archoophoran turbellarians
Subclass Neoophora	Neoophoran turbellarians

PHYLUM GNATHOSTOMULIDA — ACOELOMATE WORMS

Phylum Gnathostomulida	Acoelomate worms

PHYLUM ORTHONECTIDA

Phylum Orthonectida

PHYLUM RHOMBOZOA

Class Dicyemida
Class Heterocyemida

PHYLUM NEMERTEA — RIBBON OR PROBOSCIS WORMS

Class Anopla	Unarmed nemerteans
Class Enopla	Armed nemerteans

PHYLUM GASTROTRICHA — MARINE AND FRESHWATER GASTROTRICHS

Phylum Gastrotricha	Marine and freshwater gastrotrichs

PHYLUM NEMATODA — ROUNDWORMS

Class Adenophorea
Class Secernentea

PHYLUM NEMATOMORPHA — HORSEHAIR OR HAIRWORMS

Class Nectonematoida	Marine nematomorphs
Class Gordioida	Freshwater and semiterrestrial nematomorphs

PHYLUM ROTIFERA — ROTIFERS

Class Seisonidea
Class Bdelloidea
Class Monogononta

PHYLUM ACANTHOCEPHALA — ENDOPARASITES

Class Archiacanthocephala	Parasites of birds and mammals
Class Eoacanthocephala	Parasites of fishes and reptiles
Class Palaeacanthocephala	Parasites of all vertebrates

PHYLUM KINORHYNCHA

Order (class) Cyclorhagida
Order (class) Homalorhagida

PHYLUM LORICIFERA

Phylum Loricifera

PHYLUM TARDIGRADA   WATER BEARS

Class Heterotardigrada
Class Eutardigrada

PHYLUM SIPUNCULA   PEANUT WORMS

Phylum Sipuncula   Peanut worms

PHYLUM ECHIURA   SPOONWORMS

Phylum Echiura   Spoonworms

PHYLUM PRIAPULIDA

Phylum Priapulida

PHYLUM MOLLUSCA   MOLLUSKS

Class Aplacophora   Wormlike mollusks
Class Polyplacophora   Chitons
Class Monoplacophora
Class Gastropoda   Gastropods
Class Bivalvia   Bivalves
Class Scaphopoda
Class Cephalopoda   Squid and octopi

PHYLUM ANNELIDA   SEGMENTED WORMS

Class Polychaeta
Class Oligochaeta
Class Hrudinea   Leeches
Class Branchiobdellida

PHYLUM POGONOPHORANS

Phylum Pogonophorans

PHYLUM ARTHROPODA   ARTHROPODS

Class Insecta   Insects
Subclass Entognatha   Wingless insects
Subclass Ectognatha   Winged insects

SUBPHYLUM CHELICERATA

Class Merostomata
Class Arachnida   Spiders, scorpions, mites and ticks
Class Pycnogonida   Sea spiders

SUBPHYLUM CRUSTACEA   CRUSTACEANS

Class Remipedia
Class Cephalocarida
Class Branchiopoda
Class Ostracoda   Ostracods
Class Copepoda
Class Mystacocarida
Class Tantulocarida
Class Branchiura
Class Cirripedia   Barnacles
Class Pentastomida
Class Malacostraca   Crabs, lobsters, shrimp

SUBPHYLUM MYRIAPODA   MYRIAPODUS ARTHROPODS

Class Chilopoda   Centipedes
Class Symphyla
Class Diplopoda   Millipedes
Class Pauropoda   Pauropods

PHYLUM ONYCHOPHORA

Phylum Onychophora

PHYLUM HEMICHORDATA

Class Enteropneusta   Acorn worms
Class Pterobranchia

PHYLUM CHORDATA

SUBPHYLUM UROCHORDATA

Class Ascidiacea   Sea squirts
Class Thaliacea
Class Larvacea

SUBPHYLUM
CEPHALOCHORDATA

LANCELETS

PHYLUM CHAETOGNATHA ARROWWORMS

Phylum Chaetognatha   Arrowworms

PHYLUM
ECHINODERMATA

ECHINODERMS

SUBPHYLUM HOMALOZOA

SUBPHYLUM CRINOZOA

Class Crinoidea   Sea lilies, feather stars

SUBPHYLUM ASTEROZOA

Class Asteroidea   Sea stars
Class Ophiuroidea   Brittle stars
Class Concentricycloidea

SUBPHYLUM ECHINOZOA

Class Echinoidea   Sea urchins, sand dollars
Class Holothuroidea   Sea cucumbers

PHYLUM BRYOZOA

Class Phylactolaemata
Class Stenolaemata
Class Gymnolaemata

PHYLUM ENTOPROCTA

Phylum Entoprocta

PHYLUM PHORONIDA

Phylum Phoronida

PHYLUM BRACHIOPODA   BRACHIPODS

Class Inarticulata
Class Articulata

CLASS ARACHNIDA

ORDER ARANEAE   SPIDERS

Agelenidae	Funnel weavers
Antrodiaetidae	Folding trapdoor spiders
Araneidae	Orb weavers
Clubionidae	Sac spiders
Ctenidae	Wandering spiders
Ctenizidae	Trapdoor spiders
Dictynidae	Dictynid spiders
Dipluridae	Funnel-web spiders
Eresidae	Eresid spiders
Heteropodidae	Huntsman spiders
Linyphiidae	Dwarf spiders
Loxoscelidae	Violin spiders
Lycosidae	Wolf spiders
Oxyopidae	Lynx spiders
Pholcidae	Daddy-long-legs spiders
Pisauridae	Nursery-web spiders
Salticidae	Jumping spiders
Scytodidae	Spitting spiders
Selenopidae	Selenopid crab spiders
Sparassidae	Giant crab spiders
Tetragnathidae	Large-jawed orb weavers
Theraphosidae	Tarantulas
Theridiidae	Comb-footed spiders
Thomisidae	Crab spiders

ORDER SCORPIONES   SCORPIONS

Buthidae	Buthid scorpions
Iuridae	Iurid scorpions

ORDER
PSEUDOSCORPIONES

PSEUDO-SCORPIONS

Cheliferidae	
Chernetidae	Chernetids

ORDER OPILIONES   HARVESTMEN

Leiobunidae
Phalangiidae

ORDER ACARINA	MITES AND TICKS
Argasidae	Soft ticks
Hydrachnellae	Water mites
Ixodidae	Hard ticks
Tetranychidae	Spider mites
Trombidiidae	Velvet mites

ORDER UROPYGI	WHIPSCORPIONS
Thelyphonidae	Vinegaroons

ORDER AMBLYPYGI	TAILLESS WHIPSCORPIONS
Phrynidae	
Tarantulidae	

ORDER SOLIFUGAE	WINDSCORPIONS
Eremobatidae	
Solpugidae	

CLASS INSECTA

ORDER PROTURA	PROTURANS
Eosentomidae	

ORDER COLLEMBOLA	SPRINGTAILS
Entomobryidae	
Hypogastruridae	
Isotomidae	
Onychiuridae	
Sminthuridae	Globular springtails

ORDER DIPLURA	DIPLURANS
Campodeidae	Campodeids
Japygidae	Japygids

ORDER ARCHAEOGNATHA	BRISTLETAILS
Machilidae	Jumping bristletails
Meinertellidae	

ORDER THYSANURA	SILVERFISH
Lepismatidae	Silverfish (firebrats)

ORDER EPHEMEROPTERA	MAYFLIES
Baetidae	Small mayflies
Ephemerellidae	Midboreal mayflies
Ephemeridae	Burrowing mayflies
Heptageniidae	Stream mayflies
Leptophlebiidae	Spinners

ORDER ODONATA	DRAGONFLIES AND DAMSELFLIES
Aeschnidae	Darners
Calopterygidae	Broad-winged damselflies
Coenagrionidae	Narrow-winged damselflies
Cordulegastridae	Biddies
Corduliidae	Green-eyed skimmers
Gomphidae	Clubtails
Lestidae	Spread-winged damselflies
Libellulidae	Common skimmers
Macromiidae	Belted and river skimmers
Petaluridae	Graybacks

ORDER BLATTODEA	COCKROACHES
Blaberidae	
Blattellidae	
Blattidae	Common cockroaches

ORDER MANTODEA	MANTIDS
Empusidae	
Hymenopodidae	Flower mantids
Mantidae	Common praying mantids

ORDER ISOPTERA	TERMITES
Hodotermitidae	Rotten-wood termites
Kalotermitidae	Damp-wood termites
Mastotermitidae	
Rhinotermitidae	Subterranean termites
Termitidae	Nasutiform termites

ORDER ZORAPTERA	ZORAPTERANS
Zorotypidae	Zorapterans

ORDER GRYLLOBLATTODEA ICE INSECTS

Grylloblattidae	Rock crawlers

ORDER DERMAPTERA EARWIGS

Chelisochidae	Black earwigs
Forficulidae	Common earwigs
Labiduridae	Long-horned earwigs
Labiidae	Little earwigs

ORDER PLECOPTERA STONEFLIES

Capniidae	Small winter stoneflies
Chloroperlidae	Green stoneflies
Isoperlidae	Green-winged stoneflies
Leuctridae	Rolled-winged stoneflies
Nemouridae	Spring stoneflies
Peltoperlidae	Roachlike stoneflies
Perlidae	Common stoneflies
Perlodidae	Perlodid stoneflies
Pteronarcidae	Giant stoneflies
Taeniopterygidae	Winter stoneflies

ORDER ORTHOPTERA GRASSHOPPERS AND CRICKETS

Acrididae	Short-horned grasshoppers
Cooloolidae	Cooloola monster
Cylindrachetidae	Sand gropers
Eneopterinae	Bush crickets
Eumastacidae; Tanaoceridae	Monkey grasshoppers
Gryllacrididae; Rhaphidophoridae	Leaf-rolling crickets, camel crickets, cave crickets
Gryllidae	True crickets
Gryllotalpidae	Mole crickets
Myrmecophilidae	Ant crickets
Oecanthinae Pyrgomorphidae	Tree crickets
Stenopelmatidae	King crickets
Tetrigidae	Pygmy grasshoppers
Tettigoniidae	Long-horned grasshoppers and katydids
Tridactylidae	Pygmy mole crickets

ORDER PHASMATODEA	STICK AND LEAF INSECTS
Phasmatidae	Stick insects
Phasmidae	Walking sticks
Timemidae	Timemas

ORDER EMBIOPTERA WEBSPINNERS

Clothodidae	Clothodids
Embiidae	Embiids

ORDER PSOCOPTERA BOOKLICE AND BARKLICE

Lepidopsocidae	Scaly barklice
Liposcelididae	Liposcelid booklice
Pseudocaeciliidae	
Psocidae	Common barklice
Psyllipsocidae	Psyllipsocids
Trogiidae	Trogiid booklice

ORDER PHTHIRAPTERA PARASITIC LICE

Boopidae	
Echinophthiriidae	Spiny sucking lice
Gyropidae	Guinea pig lice
Haematopinidae	Mammal-sucking lice
Hoplopleuridae	
Laemobothriidae	Bird lice
Linognathidae	Smooth sucking lice
Menoponidae	Poultry-chewing lice
Pediculidae	Human lice
Philopteridae	Feather-chewing lice
Ricinidae	Bird lice
Trichodectidae	Mammal-chewing lice

ORDER HEMIPTERA BUGS

Achilidae	Achilid planthoppers
Adelgidae	Pine aphids
Aleyrodidae	Whiteflies
Alydidae	Broad-headed bugs
Anthocoridae	Flower bugs, minute pirate bugs
Aphididae Aphrophoridae	Aphids
Aradidae	Flat bugs, bark bugs
Asterolecaniidae	Pit scales
Belostomatidae	Giant water bugs

Berytidae	Stilt bugs
Carsidaridae	
Cercopidae	Spittlebugs and froghoppers
Chermidae	Pine and spruce aphids
Cicadellidae	Leafhoppers
Cicadidae	Cicadas
Cimicidae	Bed bugs
Cixiidae	Cixiid planthoppers
Coccidae; Colobathristidae	Soft scales, wax scales
Coreidae	Leaf-footed bugs, crusader bugs
Corixidae	Water boatmen
Cydnidae	Negro bugs
Dactylopiidae	Cochineal bugs
Delphacidae	Delphacid planthoppers
Derbidae	Derbid planthoppers
Diaspididae	Armored scale insects
Dictyopharidae; Dinidoridae	Dictyopharid planthoppers
Dipsocoridae; Schizopteridae	Jumping ground bugs
Eriosomatidae	Woolly and gall-making aphids
Eurymelidae; Membracidae	Treehoppers
Flatidae	Flatid planthoppers
Fulgoridae	Fulgorids
Gelastocoridae	Toad bugs
Gerridae	Water striders
Hebridae Homotomidae	Velvet water bugs
Hydrometridae	Water measurers
Isometopidae	Jumping tree bugs
Issidae	Issid planthoppers
Kermidae	Gall-like coccids
Kerriidae; Lacciferidae	Lac insects
Leptopodidae	Spiny shore bugs
Lygaeidae	Seed bugs
Margarodidae	Giant scale insects
Mesoveliidae	Water treaders
Miridae	Leaf or plant bugs
Nabidae	Damsel bugs
Naucoridae	Creeping water bugs
Nepidae	Waterscorpions
Notonectidae	Backswimmers
Ochteridae	Velvety shore bugs
Ortheziidae	Ensign scales
Peloridiidae	Moss bugs
Pentatomidae	Stink bugs, shield bugs
Phylloxeridae	Gall aphids
Phymatidae	Ambush bugs

Polyctenidae	Bat bugs
Pseudococcidae; Eriococcidae	Mealybugs
Psyllidae	Psyllids, lerps
Pyrrhocoridae	Red bugs, stainers
Reduviidae	Assassin bugs
Rhopalidae	Scentless plant bugs
Ricaniidae	Ricaniid planthoppers
Saldidae	Shore bugs
Scutelleridae	Jewel bugs, shield-backed bugs
Tessaratomidae	
Tettigarctidae	Hairy cicadas
Tingidae	Lace bugs
Triozidae	
Veliidae	Ripple bugs

ORDER THYSANOPTERA — THRIPS

Aeolothripidae	Banded thrips
Phloeothripidae	Tube-tailed thrips
Thripidae	Common thrips

ORDER MEGALOPTERA — ALDERFLIES AND DOBSONFLIES

Corydalidae	Dobsonflies
Sialidae	Alderflies

ORDER RAPHIDIOPTERA — SNAKEFLIES

Inocelliidae
Raphidiidae

ORDER NEUROPTERA — NET-VEINED INSECTS

Ascalaphidae	Owlflies
Chrysopidae	Green lacewings
Coniopterygidae	Dusty-wings
Hemerobiidae	Brown lacewings
Ithonidae	Moth lacewings
Mantispidae	Mantidflies
Myrmeleontidae	Antlions
Nemopteridae	
Polystoechotidae	Giant lacewings
Psychopsidae	Silky lacewings
Sisyridae	Spongeflies

ORDER COLEOPTERA — BEETLES

Anobiidae	Furniture beetles
Anthribidae	Fungus weevils
Bostrichidae	Branch and twig borers, auger beetles
Brentidae	Primitive weevils
Bruchidae	Seed beetles
Buprestidae	Metallic wood-boring beetles, jewel beetles
Cantharidae	Soldier beetles
Carabidae	Ground beetles
Cerambycidae	Longhorn beetles, longicorn beetles
Chrysomelidae	Leaf beetles
Cicindelidae	Tiger beetles
Cleridae	Checkered beetles
Coccinellidae	Ladybird beetles
Cucujidae	Flat bark beetles
Curculionidae	Snout beetles and weevils
Dermestidae	Dermestid beetles
Dytiscidae	Predacious diving beetles
Elateridae	Click beetles
Erotylidae	Pleasing fungus beetles
Gyrinidae	Whirligig beetles
Haliplidae	Crawling water beetles
Histeridae	Hister beetles
Hydrophilidae; Laemophloeidae	Water scavenger beetles
Lampyridae	Fireflies, lightning bugs
Lathridiidae	Minute brown scavenger beetles
Lucanidae	Stag beetles
Lycidae	Net-winged beetles
Lymexylidae	Ship-timber beetles
Meloidae	Blister beetles
Melyridae	Softwinged flower beetles
Mycetophagidae	Hairy fungus beetles
Nitidulidae	Nitidulid beetles
Passalidae	Passalid beetles or bessbugs
Psephenidae	Water pennies
Ptiliidae	Feather-winged beetles
Ptinidae	Spider beetles
Pyrochroidae	Fire-colored beetles
Rhipiphoridae	Rhipiphoridan beetles
Scarabaeidae	Scarab beetles
Scolytidae	Bark and ambrosia beetles
Silphidae	Carrion beetles
Silvanidae	Flat grain beetles
Staphylinidae	Rove beetles
Tenebrionidae	Darkling beetles
Trogidae	Carcass beetles
Trogositidae	Bark-gnawing beetles

ORDER STREPSIPTERA — TWISTED-WINGED PARASITES

Mengeidae	
Stylopidae	Stylopids

ORDER MECOPTERA AND RELATIVES — SCORPIONFLIES

Bittacidae	Hangingflies
Boreidae	Snow scorpionflies
Panorpidae	Common scorpionflies

ORDER SIPHONAPTERA — FLEAS

Dolichopsyllidae; Ceratophyllidae	Rodent fleas
Leptopsyllidae	Mouse fleas
Pulicidae	Common fleas
Tungidae	Sticktight and chigoe fleas

ORDER DIPTERA — FLIES

Acroceridae	Bladder flies
Agromyzidae	Leafminer flies
Anthomyiidae	Anthomyiid flies
Apioceridae	Flower-loving flies
Asilidae	Robber flies
Bibionidae	March flies
Blephariceridae	Net-winged midges
Bombyliidae	Bee flies
Braulidae	Beelice
Calliphoridae	Blowflies
Canacidae	Beach flies
Cecidomyiidae	Gall midges
Ceratopogonidae	Punkies, biting midges
Chaoboridae	Phantom midges
Chironomidae	Midges
Chloropidae	Grass flies

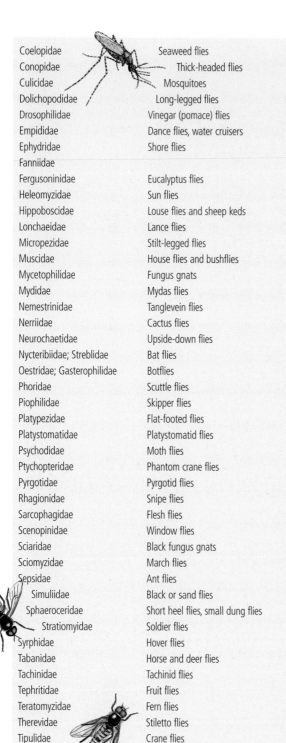

Coelopidae	Seaweed flies
Conopidae	Thick-headed flies
Culicidae	Mosquitoes
Dolichopodidae	Long-legged flies
Drosophilidae	Vinegar (pomace) flies
Empididae	Dance flies, water cruisers
Ephydridae	Shore flies
Fanniidae	
Fergusoninidae	Eucalyptus flies
Heleomyzidae	Sun flies
Hippoboscidae	Louse flies and sheep keds
Lonchaeidae	Lance flies
Micropezidae	Stilt-legged flies
Muscidae	House flies and bushflies
Mycetophilidae	Fungus gnats
Mydidae	Mydas flies
Nemestrinidae	Tanglevein flies
Nerriidae	Cactus flies
Neurochaetidae	Upside-down flies
Nycteribiidae; Streblidae	Bat flies
Oestridae; Gasterophilidae	Botflies
Phoridae	Scuttle flies
Piophilidae	Skipper flies
Platypezidae	Flat-footed flies
Platystomatidae	Platystomatid flies
Psychodidae	Moth flies
Ptychopteridae	Phantom crane flies
Pyrgotidae	Pyrgotid flies
Rhagionidae	Snipe flies
Sarcophagidae	Flesh flies
Scenopinidae	Window flies
Sciaridae	Black fungus gnats
Sciomyzidae	March flies
Sepsidae	Ant flies
Simuliidae	Black or sand flies
Sphaeroceridae	Short heel flies, small dung flies
Stratiomyidae	Soldier flies
Syrphidae	Hover flies
Tabanidae	Horse and deer flies
Tachinidae	Tachinid flies
Tephritidae	Fruit flies
Teratomyzidae	Fern flies
Therevidae	Stiletto flies
Tipulidae	Crane flies
Trichoceridae	Winter crane flies

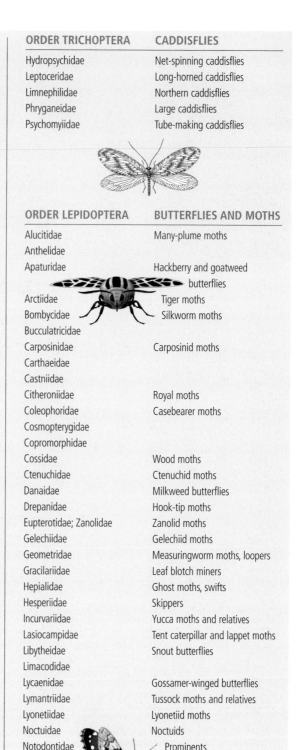

ORDER TRICHOPTERA — CADDISFLIES

Hydropsychidae	Net-spinning caddisflies
Leptoceridae	Long-horned caddisflies
Limnephilidae	Northern caddisflies
Phryganeidae	Large caddisflies
Psychomyiidae	Tube-making caddisflies

ORDER LEPIDOPTERA — BUTTERFLIES AND MOTHS

Alucitidae	Many-plume moths
Anthelidae	
Apaturidae	Hackberry and goatweed butterflies
Arctiidae	Tiger moths
Bombycidae	Silkworm moths
Bucculatricidae	
Carposinidae	Carposinid moths
Carthaeidae	
Castniidae	
Citheroniidae	Royal moths
Coleophoridae	Casebearer moths
Cosmopterygidae	
Copromorphidae	
Cossidae	Wood moths
Ctenuchidae	Ctenuchid moths
Danaidae	Milkweed butterflies
Drepanidae	Hook-tip moths
Eupterotidae; Zanolidae	Zanolid moths
Gelechiidae	Gelechiid moths
Geometridae	Measuringworm moths, loopers
Gracilariidae	Leaf blotch miners
Hepialidae	Ghost moths, swifts
Hesperiidae	Skippers
Incurvariidae	Yucca moths and relatives
Lasiocampidae	Tent caterpillar and lappet moths
Libytheidae	Snout butterflies
Limacodidae	
Lycaenidae	Gossamer-winged butterflies
Lymantriidae	Tussock moths and relatives
Lyonetiidae	Lyonetiid moths
Noctuidae	Noctuids
Notodontidae	Prominents
Nymphalidae	Brush-foots

Classifying Animals

FACT FILE

Oecophoridae	Oecophorid moths
Papilionidae	Swallowtails and relatives
Pieridae	Whites and sulfurs
Plutellidae	Diamondback moths
Pterophoridae	Plume moths
Psychidae	Bagworm moths
Pyralidae	Pyralid moths
Riodinidae	Metalmarks
Saturniidae	Giant silkworm moths

Satyridae	Satyrs, nymphs, and arctics
Sesiidae	Clear-winged moths
Sphingidae	Sphinx or hawk moths
Thaumetopoeidae	
Tineidae	Clothes moths and relatives
Tortricidae	Tortricid moths
Uraniidae	
Yponomeutidae	Ermine moths
Zygaenidae	Smoky moths

ORDER HYMENOPTERA BEES, ANTS, WASPS AND SAWFLIES

Agaonidae	Fig wasps
Andrenidae	Andrenid bees
Apidae	Digger bees, carpenter bees, cuckoo bees, bumblebees, honeybees and relatives
Argidae	Argid sawflies
Braconidae	Braconids
Cephidae	Stem sawflies
Chalcididae	Chalcids
Chrysididae	Cuckoo wasps
Cimbicidae	Cimbicid sawflies
Colletidae	Yellow-faced and plasterer bees
Cynipidae	Gall wasps
Encyrtidae	Encyrtid wasps
Eulophidae; Aphelininae	Eulophid wasps
Eurytomidae	Seed chalcids
Evaniidae	Hatchet wasps
Formicidae	Ants
Halictidae	Sweat (halictid) bees
Ibaliidae	Ibaliid wasps

Ichneumonidae	Ichneumonids
Megachilidae	Leaf-cutting bees
Melittidae	Melittid bees
Mutillidae	Velvet ants
Mymaridae	Fairy wasps
Pelecinidae	Pelecinids
Pergidae	Pergid sawflies
Pompilidae	Spider wasps
Pteromalidae	Pteromalid wasps
Scelionidae	Scelionid wasps
Scoliidae	Scoliid wasps
Siricidae	Horntails
Sphecidae	Sphecid wasps
Stephanidae	Stephanid wasps
Symphyta	Sawflies
Tenthredinidae	Common sawflies
Tiphiidae	Tiphid wasps
Torymidae	Torymid wasps
Vespidae	Vespid wasps

Size Comparisons

COMPARING MAMMALS

Blue whale

Gouldian finch

Hoopoe

Ruby-throated hummingbird

COMPARING BIRDS

Ostrich

Caribbean flamingo

King penguin

King vulture

Jungle fowl

Sulfur-breasted toucan

African elephant

Gray-headed
flying fox

Wapiti

Human

Hippopotamus

Bengal tiger

Western lowland gorilla

Red kangaroo

Nine-banded armadillo

Rat

Size Comparisons

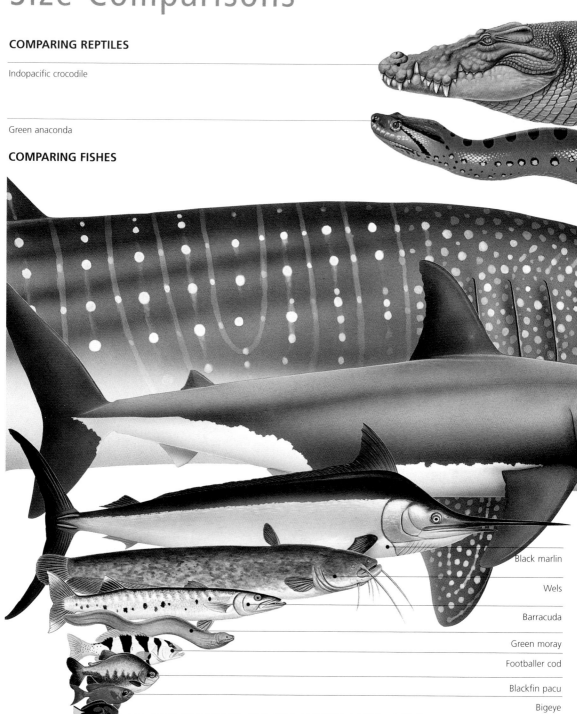

COMPARING REPTILES

Indopacific crocodile

Green anaconda

COMPARING FISHES

Black marlin

Wels

Barracuda

Green moray

Footballer cod

Blackfin pacu

Bigeye

Blue tang

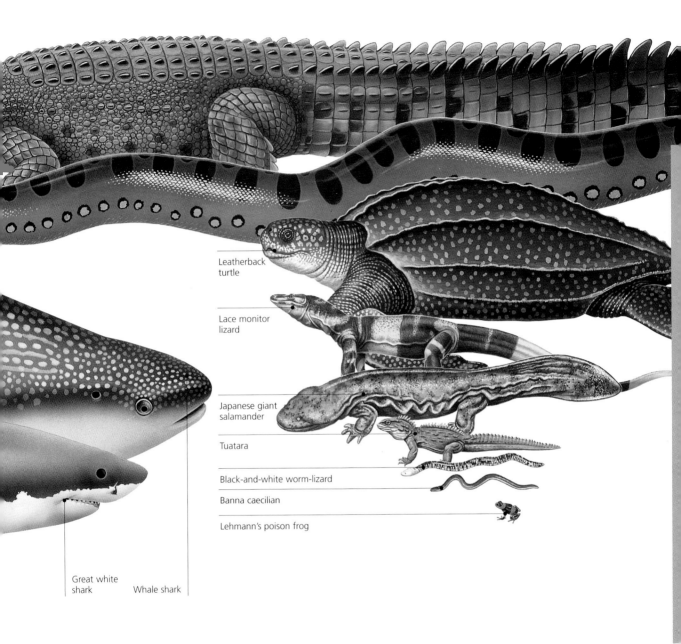

Leatherback
turtle

Lace monitor
lizard

Japanese giant
salamander

Tuatara

Black-and-white worm-lizard

Banna caecilian

Lehmann's poison frog

Great white
shark

Whale shark

Index

A

aardvark 79, 206
aardwolf 145
Ablabys taenionotus 479
Acanthaster planci 501
Acanthophis antarcticus 395
Acanthorhynchus tenuirostris 326
Acantophthalmus myersi 460
Accrididae family 524
Acinonyx jubatus 155
Acipenser
  oxyrinchus 452
  ruthenus 453
Aconaemys fuscus 227
Acrochordus
  arafurae 386
  granulatus 387
Acryllium vulturinum 277
Actitis macularia 284
adder, common death 395
Aegotheles sp. 299
Agalychnis callidryas 417
agama 366
  Iranian toad-headed 366
Agama agama 366
agouti, Brazilian 226
Ahaetulla nasuta 386
Ailuroedus dentirostris 335
Ailuropoda melanoleuca 130, 137
Ailurus fulgens 145
Aix sponsa 262
Ajaia ajaia 256
Alauda arvensis 318
albatross
  gray-headed 250
  wandering 237
Albertosaurus 57
Alcedo cristata 306
Alces alces 191
Allenopithecus nigroviridis *112*

Alligator
  mississippiensis 356
  sinesis 357
alligators 343, 356–7
Allosaurus 43, 47, 51, 57, 59
Alopex lagopus 128
Alopias vulpinus 441
Alosa sapidissima 459
Alouatta seniculus 116
alpaca 184–5
Alphadon 66
alpine regions 26–7
Amblyornis macgregoriae 335
Amblyrhynchus cristatus 365
Ambystoma
  laterale 404
  mexicanum 406
  tigrinum 405
Ameiva ameiva 379
Amia calva 452
ammonite 39
amphibians
  classifying 400–1, 563
amphisbaenians 360–1
Amphiuma means 406
Anableps anableps 475
anaconda 582–3
Anas
  cyanoptera 263
  rubripes 263
Anatotitan 60
anchoveta, Peruvian 458
Andrias
  alpestris 407
  japonicus 407
anemonefish, spinecheek 483
angler, Paxton's whipnose 471
Anguilla rostrata 456
Anilius scytale 383
anole
  Cuban brown 365
  dewlap 362
Anolis sagrei sagrei 365

Anopheles sp. 538
Anostomus anostomus 463
Anser anser 262
anteater, giant 93
antelopes 193–5
  four-horned 195
  sable 194
  saiga 195
antennae 519
Anthus spinoletta 318
Antilocapra americana 187
Antilope cervicapra 195
antlers 188–9
ants 546–9
  Amazon 549
  bulldog 546
  leaf-cutter 549
  red 549
Aotus trivirgatus 116
apalis, bar-throated 314
Apaloderma narina 304
Apalone spinifera spinifera 353
apes 118–21 *see also* monkeys, primates
Aphelocoma coerulescens 337
Aphredoderus sayanus 470
Aphyosemion gardneri 475
Apidae family 546
Aplocheilus annulatus 475
Apus affinis 302
aracari, curl-crested 313
Aramus guarauna 278
arawana 454
Archaeopteryx 43, 66, 72–3, 240
Archelon 67, 347
Archilochus
  alexandri 302
  colubris 302
Architeuthis sp. 496
Arctiidae family 541
Arctocebus calabarensis 108
Arctonyx collaris 139

capybara 216, 226
Carabidae family 533
caracal 155
caracara
    crested 264, 271
    striated 270
Caracara plancus 270
Carcharhinus melanopterus 438
Carcharias taurus 441
Carcharodon carcharias 441
Carcharodontosaurus 56
cardinals 329
Carduelis
    carduelis 331
    chloris 331
caribou 188, 190
Carnegiella strigata 463
carnivores 122–3, 142–3
cassowary 245
Castor
    canadensis 222
    fiber 223
Catagonus wagneri 181
caterpillars 541, 545
catfish 460–3
    electric 461
    striped sea 426, 462
cats 148–57
    African golden 157
    Andean mountain 156
    Asiatic golden 148, 157
    fishing 156
    jungle 156
    Pallas' 148
cattle 23, 178, 198–9
Cavia aperea 227
Cebus apella 116
centipedes 491, 503, 514–17
    giant desert 514
Centrophorus moluccensis 436
Centrosaurus 65
Centurio senex 102
century plant flowers 21

Cephalophus sp. 194
Cephalorhynchus
    commersonii 175
    hectori 174
Cerambycidae family 532
Ceratophrys ornata 415
ceratopians 64–5
Ceratosaurus 43
Ceratotherium simum 201
Cercopithecus diana 113
Cerocebus agilis 112
Cervus
    duvaucelii 190
    elaphus canadensis 191
Cestoidea class 498
Cetorhinus maximus 441
Chaetodipus formosus 225
Chaetodon meyeri 481
chain-fruit cholla 21
Chamaeleo
    jacksonii 365
    lateralis 367
chameleons 366
    Jackson's 362, 365
    Knysna dwarf 367
    Malagasy 367
characin, flying 427
Charadrius vociferus 284
Chasmosaurus 64
Chauliodus macouni 469
Chauna chavaria 262
cheetah 148, 155
Cheirolepis 431
chimaeras 444–5
    blunt-nosed 446
    shortnose 445
chimpanzee 104, 119
chinchilla 227
chipmunk
    eastern 219
    least 217, 220
Chirocentrus dorab 459
Chironectes minimus 88

Chitala chitala 455
Chlamydosaurus knigii 367
Chlamydoselachus anguineus 443
Chloebia gouldiae 330
Chlorocebus aethiops 113
Chloropsis
    aurofrons 321
    hardwickii 321
Choerodon fasciatus 482
Chondrohierax urcinatus 267
chordates 492–3
chough 336
Chrysemys picta belli 352
Chrysococcyx cupreus 294
Chrysocyon brachyurus 127
Chrysomelidae family 532
cicada 502, 518, 526–7
cichlid, Lake Malawi zebra 483
Cicinnurus regius 334
Cidaris cidaris 500
Cinnyris regius 327
Cisco 464
Cissa chinensis 332
civet, African 122, 123, 142–3
Civettictis civetta 143
Cladoselache 431
classification 78–9, 236–7, 342–3,
        400–1, 490–1, 554–81
claws 85, 107, 148, 265
Cleidopus gloriamaris 477
Clethrionomys glareolus 225
Climatius 430
clingfish 474–5
clouded leopard 152
Clupeonella cultriventris 458
Cnemidophorus
    lemniscatus 377
    tesselatus 376
cnidarians 492–3
coati, South American 145
cobras 142, 390–1
    Egyptian 380, 392
    spitting 362, 390
Coccinellidae family 534

G

hogs 181
Homaloptera orthogoniata 461
Homoroselaps lacteus 388
honeycreeper, red-legged 328
honeyeater
    banded 314
    regent 326
hoopoe 307, 580
hornbill, yellow-billed 309
horns 178
horses 79, 178, 202–5
horsetail fern 14
hound, pharoah 123
hummingbirds 236, 300–3
    Anna's 301, 303
    black-chinned 301, 302
    blue-throated 303
    broad-tailed 303
    Calliope 303
    Costa's 301, 303
    ruby-throated 301, 302, 580
    rufous 301, 302
hunting 128, 166, 253
Huso huso 453
Hydrochaeris hydrochaeris 226
Hydrolaetare schmidti 415
Hydrolagus colliei 446
Hydrosaurus ambionensis 366
Hyemoschus aquaticus 190
hyena 146–7
    spotted 122, 123, 144, 148
    striped 144
Hylinobatrachium fleischmanni 417
Hylonomus 38, 346
Hymenopodidae family 521
Hyperolius marmoratus 421
Hyperoodon ampullatus 173
Hypsignathus monstrosus 100
Hypsilophodon 46
Hypsipyrmnodon moschatus 90
hyraxes 27, 79, 206
    rock 27, 206
    yellow-spotted 206

Hystix
    brachyura 226
    cristata 226

I

Ichneumonidae family 548
Ichthyophis
    bannanicus 409
    kohtaoensis 409
Ichthyornis 240
Ichthyosaurus 67, 346–7
Ichthyostega 38
Icterus
    auratus 333
    bullockii 332
    nigrogularis 333
Idiacanthus fasciola 469
iguana 368–9
    Fijian crested 368
    helmeted 364
    marine 342, 362, 365, 369
iguanids 364–5
Iguanodon 61
Imantodes cenchoa 388
indri 106
Indricotherium 83
Inia geoffrensis 176
invertebrates
    classifying 490–1, 573–81
Isistius brasiliensis 436
Isurus oxyrinchus 441

J

jacamar, rufous-tailed 313
jackal, golden 128
jackdaw, Eurasian 337
Jaculus sp. 225

jacuna 283
jaguar 148, 153
jaguarundi 157
jays 336–7
jellyfish 38, 491, 493
jerboa 216, 225
julie, striped
jungle runner 379, 580

K

kagu 281
kakapo 291
Kaloula pulchra 420
kangaroos 87, 90–1
    goodfellow's tree 90
    musky rat 90
    red 87, 581
    western grey 90
Kannemeyeria 41
Kassina senegalansis 420
kea 291
Kentropix calcarata 377
kestrel
    American 270
    greater 270
    lesser 264
kiang 205
killdeer 284
killifish
    clown 472, 475
    steel-blue 475
kingbirds 316–17
kingfishers 306–9
    belted 308
    malachite 306
    white-throated 236, 306
kinglet, golden-crowned 325
kinkajou 144
kiskadee, great 236, 314, 316

R

rabbits 78, 228–31
  antelope jackrabbit 229
  black-tailed jackrabbit 229
  brush 230
  pygmy 230
  rivenne 230
  ryuku 230
  volcano 230
raccoons 122, 123, 130,
    144–5
rail
  buff-banded 280
  Woodford's 278
rainbow lorikeet 291, 292
rainbowfish
  Australian 473
  Madagscar 472, 475
rainforests, tropical 16–17
Raja sp. 446
Rallina tricolor 278
Ramphastos
  sulfuratos 312
  toco 310
Ramphotyphlops braminus 382
Rana sp. 418
  palustris 419
Rangifer tarandus 190
raptors 264–73
rats 216, 581
  black 225
  crested 224
  desert kangaroo 216, 225
  house 225
  spiny 227
rattlesnakes 390–1
  lance-headed 395
Rattus
  mindorensis 225
  rattus 225
Raturfa bicolor 220
raven 337

rays 444–7
  Atlantic devil 445
  blind electric 446
  electric 434, 445
  manta 446
  sting *see* stingrays
  torpedo 445
Regalecus glesne 476
Regulus
  ignicapillus 324
  regulus 324
  satrapa 325
reindeer 179
remora 480
reproduction 36–7, 80, 86, 132,
    348, 435, 448, 479, 506
reptiles
  characteristics 344–5
  classifying 342–3, 561–2
  evolution 346–7
resplendent quetzals 236, 304,
    305
Rhacophorus nigropalmatus 421
Rhamphorynchus 43
rhea 245
Rheobatrachus silus 414
Rhincodon typus 443
Rhinoceros
  sondaicus 201
  unicornis 201
Rhinoderma darwinii 421
Rhinolophus sp. 102
Rhinomuraena quaesita 456
Rhinophis drummondhayi 383
Rhinophrynus dorsalis 412
rhinoceros 79, 200–1
  white 178–9
Rhodeus sericeus 460
Rhyacotriton olympicus 406
Rhynchocyon chrysopygus 215
Rhynchops niger 285
Rhynochetos jubatus 281
ribbonfish 477

ricefish, Javanese 475
ringtail 145
robin 315, 322–4
rockfish, tiger 472, 479
rodents 216–27
roller, lilac-breasted 309
Romerolagus diazi 230
rosella, Eastern 291, 293
Rostrhamus sociabilis 267
roughy, Southern 472, 477
Rousettus egyptiacus 101
Rupicola rupicola 317

S

Sagittarius serpentarius 272
Saguinus
  midas 116
  oedipus 117
Saiga tatarica 195
salamanders 402–7
  blue-spotted 404
  fire 402–4
  Fischer's clawed 401, 402, 404
  Japanese giant 407, 583
  Olympic torrent 401, 402, 406
  Pacific giant 401, 402
  red 401, 402, 407
  tiger 401, 402, 405
  two-lined 406
Salamandra salamandra 404
Salmo salar 467
salmon 464–7
  Atlantic 426, 467
  sockeye 464–5, 467
Saltasaurus 45, 47
Saltopus 41
Salvelinus namaycush 466
sandgrouse 288–9
sandpiper, spotted 283, 284
Sanopus splendidus 470

stingrays 444–7
  ocelated freshwater 446
  sixgill 446
Stomiidae family 468–9
Stomiiformes order 468
stork 343
  painted 236, 257
  saddle-billed 254
strawberry 15
Streptopelia turtur 288
Strix occidentalis 297
Struthiomimus 46, 56
sturgeon 448, 452
Sturnus vulgaris 333
Sturt's pea 18
Styrachosaurus 49, 64
sunbird, regal 327
sunbittern 279
Suricata suricatta 143
Sus
  barbatus 181
  scrofa 181
swallows 237, 318–19
swans 261, 263
swifts 301–3
  Alpine 303
  little 302
swordtail 474
Sylvia borin 325
Sylvilagus bachmani 230
Syncerus caffer 198
Synchiropus splendidus 474
Synodus variegatus 468
Syrmaticus reevesii 277

T

Tabanidae family 536
Tachybaptus ruficollis 249
Tachyglossus aculeatus 84
Tachymarptis melba 303

tahr, Himalayan 199
tails 54, 62–3, 125, 163, 363,
    429, 438, 444
takahe 280
Talpa europaea 95
tamandua 92
tamarin
  cotton-top 105, 117
  midas 116
Tamias
  minimus 220
  striatus 219
Tamiasciurus
  douglasii 219
  hudsonicus 220
tanagers 328–9
Tangara fastuosa 328
tapeworms 491, 498
tapirs 178, 206–7'
  Baird's 207
  Malayan 207
  mountain 207
  South American 207
Tapirus
  bairdii 207
  indicus 207
  pinchaque 207
  terrestris 207
tarantula 491, 504, 506
Tarentola mauritanica 373
tarpan 205
tarsier 104, 106
  Philippine 109
  spectral 108
Tarsius sp. 104, 106
  spectrum 108
  syrichta 109
Tasmanian devil 87
Tauraco leucotis 294
Taxidea taxus 138
teal, cinnamon 263
teeth 49, 56, 107, 122, 170,
    355, 390, 435, 507, 509

temperate regions 22–3
temperature, regulation of 345
tenrec 95
  streaked 95
Tephriditidae family 539
Terathopius escaudatus 269
Teratoscincus scincus 371
Teratornis merriami 241
termites 522–3
terns 282–3, 287
Terrapene carolina carolina 352
terror-bird 240
Tetracerus quadricornis 195
Tetraodon biocellatus 484
Tettigoniidae family 524
Thamnophis sauritus sauritis 388
Thelotornis capensis 386
Theropithecus gelada 112
theropods 56–9
thornbill, purple-backed 300
thorny devil 366
thrush, scaly 314, 322
Thymallus thymallus 466
tick 503, 512–13
tiger 79, 123, 148, 151, 581
Tiliqua sp. 378
  scincoides 376
tinamou 236, 245
titi, masked 117
titmouse, plain 327
tits 314, 236–7
toadfish, splendid 470
toads 400, 410–21
  Asian horned 400, 410, 412
  Asiatic climbing 417
  crucifix 414
  Eastern narrow-mouthed 421
  European green 417
  Everett's Asian tree 416
  leopard 416
  natterjack 400
  oriental firebelly 400, 410, 413
  ornate horned 415

Acknowledgments

Weldon Owen would like to thank the following people for their assistance in the production of this book: Philippa Findlay and Sarah Plant at Puddingburn Editorial Services, Andrew Davies at Creative Communications, Brendan Cotter, Helen Flint, Grace Newell and Guy Troughton.

The following illustrations are © MagicGroup s.r.o. (Czech Republic) www.magicgroup.cz:
(t=top, tl=top left, tc=top center, tr=top right, cl=center left, c=center, cr=center right, b=bottom, bl=bottom left, bc=bottom center, br=bottom right).

78tc, 79tr, 92bl, 94tr, 102tr, 106t,cl, 108tl,bl,br, 109t,tr,br, 111cl,cr, 112c,bl, 113tl,cr,bl,br, 114t,cl, 115tr,bl, 116t,cl, 117br, 138t,bl,br, 139tr,cl,cr, 140c,b, 141tl,tr,cl,cr,bl,bc,br, 159tc, 160tr,c,bl,br, 161tl,tr, 178tc, 180b, 181bl,br, 186b, 190cr, 201c, 205tc, 206tr,c,bl, 207tl,tr,c,bc, 212c,b, 213t,b, 215tl,tr,cl,br,bl, 216tl,tc, 217c,br, 223tl, 224cl,c,cr, 225c,bl,br, 226br, 227tl, 230tr,cl,c,cr,bl,br, 231tl,tr, 237tr, 249cl,br, 250cl, 270br, 274tr, 276c, 278tr,cl,cr, 281bl, 291br, 292cl,bl, 293br, 294bl,bc,br, 295br, 298tr,cl,c,br, 299tr,cl, 301tc, 302tr,br, 303cl, 304c, 305tl,tr,c, 307tl,cr, 308tr,cr, 311cl, 313tl,c,bl,br, 314t, 320cl, 321bl,br, 332br, 334cl,bl,br, 366bl, 368tr,cl,cr, 370b, 371b, 373b, 374c, 375t, 377br, 378cl, 382cl, 383t,tr,bl, 392cl, 400tl,tc, 401tc,tr, 402tr, 404cl, 405c, 406tr, 407b, 408cl, 409cl,bl,br, 410tl,tr, 412tr, 413cl,c, 414t,cl,br, 415br, 416tr, 417tr,br, 426tc,br, 427tl,tc,tr,bl,bc, 432tr,c,bc, 448tl,tc,tr, 452tr,c, 453tc,c, 454tr,bc, 455bc, 458tr,cl,cr,bl,br, 459tc,c, 460tr,cl,bc,br, 461bc, 464tr,bc, 466tl,tr,c,bc, 467bc, 469tl, 470tr, 471bl,br, 476cl, 491tl, 493tl,tc,tr, 496cl, 500bl,br, 501tl, 514tl,tr,bl, 565cr, 566tc, 567cr, 568tr,c

Illustrators: Susanna Addario, Art Studio, Mike Atkinson/Garden Studio, Paul Bachem, Alistair Barnard, Priscilla Barret/Wildlife Art Ltd., Jane Beatson, Sally Beech, Andre Boos, Anne Bowman, Peter Bull, Martin Camm, Stuart Carter/Wildlife Art Ltd., D. Cole/Wildlife Art Ltd, Marjorie Crosby-Fairall, B. Croucher/Wildlife Art Ltd., Marc Dando/Wildlife Art Ltd., Peter David, Fiammetta Dogi, Sandra Doyle/Wildlife Art Ltd., Gerald Driessen, Simone End, Christer Erikson, Alan Ewart, Lloyd Foye, John Francis/Bernard Thornton Artists UK, Peg Gerrity, John Gittoes, Mike Golding, Mike Gorman, Ray Grinaway, Gino Hasler, Tim Hayward/ Bernard Thornton Artists UK, Dr. Stephen Hutchinson, Robert Hynes, Ian Jackson/Wildlife Art Ltd., Janet Jones, Roger Kent, David Kirshner, Frank Knight, Angela Lober, Frits Jan Maas, John Mac/FOLIO, David Mackay, Martin Macrae/FOLIO, Rob Mancini, Karel Mauer, Iain McKellar, James McKinnon, David Moore/Linden Artists, Robert Morton, Colin Newman/Bernard Communications, Ken Oliver/Wildlife Art Ltd., Erik van Ommen, Nicola Oram, Photodisc, Maurice Pledger, Tony Pyrzakowski, Oliver Rennert, John Richards, Edwina Riddell, Steve Roberts/Wildlife Art Ltd., Barbara Rodanska, Trevor Ruth, Claudia Saraceni, R.T. Sauey, Michael Saunders, Peter Schouten, Peter Scott/Wildlife Art Ltd., Rod Scott, Chris Shields/Wildlife Art Ltd., Ray Sim, Marco Sparaciari, Chris Stead, Kevin Stead, Mark Stewart, Stockbyte, Roger Swainston, Steve Trevaskis, Thomas Trojer, Guy Troughton, Chris Turnbull/Wildlife Art Ltd., Glen Vause, Genevieve Wallace, Trevor Weekes, Rod Westblade, Ann Winterbotham.